D1030349

162886

DATE DUE

| | | | |
|---|---|---|---|
| | | | |
| | | | |
| | | | |
| | | | |
| | | | |
| | | | |
| | | | |
| | | | |
| | | | |
| | | | |
| | | | |
| | | | |

THE MAKING OF GEOLOGY

# THE MAKING OF GEOLOGY

## Earth science in Britain
## 1660–1815

ROY PORTER

*Director of Studies in History*
*Churchill College, Cambridge*

CAMBRIDGE UNIVERSITY PRESS

. CAMBRIDGE

LONDON · NEW YORK · MELBOURNE

Published by the Syndics of the Cambridge University Press
The Pitt Building, Trumpington Street, Cambridge CB2 IRP
Bentley House, 200 Euston Road, London NW1 2DB
32 East 57th Street, New York, NY 10022, USA
296 Beaconsfield Parade, Middle Park, Melbourne 3206, Australia

First published 1977

Printed in Great Britain by
Western Printing Services, Ltd, Bristol

*Library of Congress Cataloguing in Publication Data*
Porter, Roy, 1946–
The making of geology.
Bibliography: p.
Includes index.
1. Geology – Great Britain – History. I. Title.
QE13.G7P67   550'.941   76–56220
ISBN 0 521 21521 8

*To my parents*

# Contents

# Preface

In writing this book, I have benefited from the help, and received the advice, of far more people than can individually be mentioned here. I must, however, single out debts of gratitude where these are greatest. To the constant encouragement, criticism and example of Professor J. H. Plumb and Quentin Skinner I owe so much of my training and outlook as a historian. From 1967 Robert M. Young fired me with an enthusiasm for a history of science which would come to grips with the central concerns of human history. Martin Rudwick in part supervised my Ph.D. research, and has since then been a constant source of expert help and criticism in the difficult business of transforming a thesis into a book. He has helped me to appreciate the intrinsic patterns and problems of the practice of geology, and by his own scientific expertise has saved me from countless technical errors.

Many people have read parts or all of this work in its various stages, and I should particularly like to thank Dr V. A. and Mrs J. M. Eyles, Dr Charles Webster, Dr Hugh Torrens and Mr Jack Morrell for detailed comments, correcting errors and offering bibliographical advice. Over the last few years, Bill Bynum, Ludmilla Jordanova and Mike Neve have been constant sources of friendly personal encouragement and general historical stimulus. Responsibility for the overall interpretation and for remaining errors must, however, rest firmly with myself alone.

I should like to thank the following institutions for allowing me to quote from unpublished materials in their possession: the Bodleian Library, Oxford; Bristol Central Library; the British Library; Cambridge University Library; Edinburgh University Library; the Literary and Philosophical Society of Newcastle-upon-Tyne; the National Library of Scotland; the Royal Geological Society of Cornwall; the Royal Society of London; the Royal Society of Edinburgh. I should like to thank the staffs of these and other institutions for their courteous help and advice throughout work on this book.

Much of the original research was made possible by the kindness of the Master and Fellows of Christ's College, Cambridge, in electing me to a research fellowship in 1970. I have subsequently been able to write up this book in the friendly and sympathetic atmosphere of Churchill College, Cambridge.

*June 1976*                                        ROY PORTER

Which Subject, if we consider as it is thus represented, doth look very like an Impossibility to be undertaken even by the whole World, to be gone through within an Age, much less to be undertaken by any particular Society, or a small number of Men. The number of Natural Histories, Observations, Experiments, Calculations, Comparisons, Deducations and Demonstrations necessary thereunto, seeming to be incomprehensive and numberless: And therefore a vain Attempt, and not to be thought of till after some Ages past in making Collections of Materials for so great a Building, and the employing a vast number of Hands in making this Preparation; and those of several sorts, such as Readers of History, Criticks, Rangers and Namesetters of Things, Observers and Watchers of several Appearances, and Progressions of Natural Operations and Perfections, Collectors of curious Productions, Experimenters and Examiners of Things by several Means and several Methods and Instruments, as by Fire, by Frost, by Menstruums, by Mixtures, by Digestions, Putrefactions, Fermentations, and Petrifactions, by Grindings, Brusings, Weighings and Measuring, Pressing and Condensing, Dilating and Expanding, Dissecting, Separating and Dividing, Sifting and Treining; by viewing with Glasses and Microscopes, Smelling, Tasting, Feeling, and various other ways of Torturing and Wracking of Natural Bodies, to find out the Truth or the real Effect as it is in its Constitutions or State of Being.

(ROBERT HOOKE, 1705: 279)

No Stone hath been left unturned.

(J. WOODWARD, 1695: 44)

Indeed, if the face of the earth were divided into districts, and accurately described we have no doubt that, from the comparison of these descriptions, the true theory of the earth would spontaneously emerge without any effort of genius or invention.

([J. PLAYFAIR], 1811: 209)

If only the Geologists would let me alone, I could do very well, but those dreadful Hammers! I hear the clink of them at the end of every cadence of the Bible verses.

(JOHN RUSKIN, letter of 24 May 1851 to HENRY ACLAND.
*The works of John Ruskin*, xxxvi: 115.)

# Introduction

... A species of mental derangement, in which the patient raved continually of comets, deluges, volcanos and earthquakes; or talked of reclaiming the great wastes of the chaos, and converting them into a terraqueous and habitable globe. This unreal mockery, however, though it has endured long, and continued even to the present day, is now vanishing and melting into air. ([Playfair], 1811: 207–8)

John Playfair's dismissal of previous geological theories typifies the root-and-branch denial by most early nineteenth-century geologists of their scientific parentage. The path towards the true science of the Earth, they thought, would not merely be long and arduous. Rather, they needed to strike out completely afresh, on the new road of painstaking observation and disciplined induction. For earlier theories had been jerry-built on illegitimate intellectual foundations and vitiated by rampant armchair speculation. Above all, previous systems were no fit foundation for the fast-rising edifice of geological facts. As John Kidd, Professor of Chemistry at Oxford, concluded:

Considering then of how very small a portion even of the earth's surface we have at present any thing like an accurate knowledge; ...considering also, that whoever has long accustomed himself to geological observations, will easily recollect that at different periods he has viewed the same phenomena with very different eyes; and that the history of geology shews the same thing to have happened to the most acute and accurate observers; that it often also happens that different persons are impressed differently at the same moment by the same phenomena; and lastly, that it is certain from the numerous and remarkably dissimilar systems of different philosophers, that nothing like probability of any high order has been yet attained in geological reasoning; from all these considerations we may at least be convinced, that the science of geology is at present so completely in its infancy as to render hopeless any attempt at successful generalization, and may therefore be induced to persevere with patience in the accumulation of useful facts. (1815: 268–9)

George Bellas Greenough, first President of the Geological Society of London, capped Kidd's scepticism in his own *Critical examination* (1819). Abelard-like, he juxtaposed theories to spotlight their contradictions. But he also exposed how all 'facts' and categories were themselves theory-bound: 'The term Stratification is by no means unconnected with theory' (1819a: 23). Like Kidd, Greenough sought to wipe the slate clean with an intensive campaign of observation and data collection. In the Geological Society's Baconian crusade he was in the van.[1] So fervent was this new repugnance to 'speculation' that not only all hitherto-existing theories, but also the very activity of theorizing seemed ripe for the holocaust, now that geology had a

tendency in all its branches, to assume a character of strict experiment or observation, at the expense of all hypothesis, and even of moderate theoretical speculation...Matter of fact methods have lately been gaining ground in Geology, as in other sciences; hypotheses are now scarcely listened to; and even the well-organized theories which, a short time since, created so much controversy, receive in this day little attention or comment. ([Fitton], 1817a: 175, 177)

Denunciation of preceding theories was two-pronged. They were religiously motivated, and, therefore, distorted; and their science was merely speculative. The charge that geology had hitherto been in thrall to Moses reached its apogee in Charles Lyell's polemical historical introduction to his *Principles of geology* (1830, i), in which he scornfully dismissed 'the familiar association in the minds of philosophers, in the age of Newton, of questions in physics and divinity' (p. 36).[2] The upshot, Lyell lamented, was that the science's history over the last two centuries had been one of 'retardation as well as advance', so that his 'inquiry, therefore,...is singularly barren of instruction to him who searches for truths in physical science' (p. 30). As late as 1830 Lyell's 'mission' was to 'free the science from Moses' (Mrs Lyell, 1881, i: 268). The latter charge was no less pressed, that earlier geologists had shirked building up their explanations by sober induction from patient investigation, leaping instead to pretended causes filched from other disciplines (or, worse, from none at all). For the problem of relating past and present states of the Earth had fired the imagination of some speculative philosophers better versed in mathematics, astronomy, and geography, than in chymical or mineralogical knowledge, yet all desirous of tracing the origin of the globe and applying Geology to their several systems. Thus COSMOGONY was grafted on Geology. (Kirwan, 1799: v)

That geology was 'not to be confounded with cosmogony' became of course axiomatic (Lyell, 1830, i: 1).

Standard historiography has remained loyal to this assessment of the science's early history. It is amused by the cosmogonists, and almost ignores the fieldwork, of the seventeenth century. It then stigmatizes the bulk of the eighteenth century as a dark age, and attributes the real paternity of British geologists to heroes such as James Hutton, William Smith and Charles Lyell. Furthermore, it makes scant recognition of the profounder issue of the very formation of geology itself as a science.

Contrary to these early nineteenth-century geologists from whom most historians have taken their cue, I wish to argue that attitudes towards the Earth and its investigation underwent great transformation in Britain between the mid-seventeenth and the early nineteenth century.[3] The labours of the seventeenth and eighteenth centuries provided the basis – material and conceptual – for the unquestioned flowering of geology in the nineteenth, not, as was claimed, an obstacle to be demolished before progress could be made.

My aim here, of course, is not primarily to redistribute glory, but rather to reconceptualize what was involved in the origins and growth of the science. For this transformation *is* the emergence of *geological* science in Britain, 'the making of the science of geology'. To some extent, this process was tangible, material and quantifiable: growing numbers of observers and observations, maps and monographs, collections and societies. But it was also broader and deeper. During this period Earth science was gaining a *social* base hitherto unknown. Being created, slowly and incoherently, was a highly complex institutional and intellectual fabric of devotees, facts, ideas, ambitions and controls, which mid-seventeenth-century pioneers like Hooke had been sorely aware were lacking (1705: 279), but which had fused into a distinctive, self-sustaining discipline by the early nineteenth.

But investigation of the Earth also underwent parallel *conceptual* change. There was continuous pressure to render the pursuit more 'scientific', while criteria of the 'scientific' themselves developed in course of time. Methods, techniques and standards were forged which were claimed to be more rigorous, philosophically sophisticated and appropriate to the object. Pragmatically speaking, these moves bore fruit. But to describe the investigation of the Earth as becoming more 'scientific' is also to imply that something like a coherent public discipline was being built up, with its own distinctively demarcated intellectual territory, construction of reality, and its own practices and practitioners.

My historical perspective, in other words, sees science as an integral but distinct part of the spectrum of man's intellectual and social activities. Human thought is like white light. When it passes through the prism of the outside world, it is fragmented into bands of colours, continuous but recognizably different. Science – like poetry or religion – is one of those colour bands. The historian must embrace the *historical* fact that natural sciences have shown a tendency, when contrasted with many other forms of discourse, to be descriptively *cumulative*, in the sense that new knowledge is successively acquired which can be used alongside other data. Science, likewise descriptively considered, has a tendency to be *progressive*, in the sense that inadequate theories have successively been discarded; *objective*, in that scientists have tended to operate within common traditions and shared public languages; and *rational*, in that in scientific culture, beliefs have been backed with validating reasons (Merton, 1967). Scientists – and, indeed, philosophers of science – have frequently, but fallaciously, taken for granted such claimed features of science as openness, progress, objectivity and impersonality, as unproblematically true. But this is to ignore the historical dimension. The historian's task, rather, is to see these very aspects of the culture of science as his *problem*. For to soft-pedal these historical features of the activity of science would shallowly conflate and reduce it to the history of ideas, of ideology, or the sociology of the professions. From the historian's point of view, the question of possible ontological reasons for natural science's distinctiveness as an activity is unanswerable and irrelevant. But he must not neglect science's special historical characteristics (cf. Rudwick, 1975). For a comment on these historiographical presuppositions, cf. the Bibliographical Note, p. 237.

This poses the question: the science of *what*? At the beginning of my period, men saw the Earth through various heterogeneous traditions of discourse. By its end, some practised 'geology'. Changing ways of seeing are signposted by shifts in terminology – the coining of the word 'geology' and its derivatives.[4] This process indicates that the scientific questions which were asked about the Earth, and the conceptualization of the Earth itself, changed dramatically over a century and a half. To speak very baldly, investigating individual terrestrial products and features as isolated objects, perhaps within a philosophy of Creation, gave way to considering the Earth as a fully articulated, historically-related system of forces and materials. Static natural history gradually yielded to a dynamic 'directionalist' programme, concerned to relate the Earth's past and present, and to understand each in terms of the other. This shift was fundamental, marking the moment of the first historical science, with all its attendant functions and problems. William Whewell deftly acknowledged this in coining the term 'palaetiological' to define the historical sciences, of which geology was 'the representative' (1837, iii: 481).

My aim then is to explain what is meant by saying that investigation of the Earth became 'scientific' and 'geological'. I shall concentrate on the manner in which this occurred – that the science was *made*. This notion is in part suggested by Ravetz's historically fruitful conception of the scientist's work as a 'craft activity', which emphasizes the scientist as agent, not merely as thinker, without minimizing that what the scientist does, is intellectual:

1. Scientific inquiry is a craft.
2. The objects of this work are not natural things, but are intellectual constructs, studied through the investigation of problems.
3. The work is guided and controlled by methods which are mainly informal and tacit, rather than public and explicit.
4. The special character of achieved scientific knowledge is explained by the complex social processes of selection and transformation of the results of research. (1971: 71–2)

The man of science cannot breathe *in vacuo*. His work is to be grasped in a social environment; though not primarily in respect of external 'influences', but mediated through the shared norms, goals, standards and practices of the community, however defined, within which he is working.

Science involves the generation of 'knowledge' by human selection. The historian will investigate the selection mechanisms, the resultant science, and how the former determine the latter. A proper historical view, therefore, of a particular science's configuration at any time will not home in on those scientific ideas with most affinity to present beliefs. It will present a comprehensive vista of a constellation of scientific theories – in harmony or in conflict – facts, suppositions, metaphysical and methological surmises; facts and theories generally regarded as exploded; facts and theories yet to come into prominence and respectability, and which are (so to speak) candidates for recognition. And it will take account of the physical dimensions of the science – museums, laboratories, journals, experimental techniques and instruments. Deep historical research has invariably demonstrated the multiplicity, complexity and even incoherence of all those bids for knowledge which at any time make up the activity of a science. Such a 'truly historical approach to the history of science' would consider science as

the ensemble of activities of the scientist in the pursuit of his goal of scientific observation and understanding. It includes the various influences that affect him significantly, perhaps unknown to himself, in this pursuit. It contains all the propositional formulations, both provisional and 'finished', with the reasonings *actually* followed (not just those ultimately reported). In short, . . .everything the scientist actually *does* that affects the scientific outcome in any way. (McMullin, 1970: 16)

At this level, then, the historian's task is to explain why some parts of this body of thought, data and speculation continue in currency, while others develop, or are transformed, or disappear, temporarily or permanently.

All sciences are *made*. They are fabrics constructed by human choice and work. Sciences hinge upon facts, but facts themselves are *facta*, factitious. There is, after all, nothing foregone and eternally right in investigating the Earth *geologically*. Geology is simply the way in which modern European civilization has opted to explore it.[5] Geology has been, indeed, in some respects more conspicuously *made* than many other sciences. It is recent. It relies upon other, antecedent, sciences.[6] Furthermore, geology is exceptionally dependent upon deep social and institutional foundations. It needs a broad army of workers bestowing abundant data from diverse terrains. Hence, this book's aim is to show how something like a 'critical mass' of those interested in the Earth grew up, forging the necessary techniques, conceptions of Nature, critical standards, collections, publications and such like; none of which singly, but all of which together created the science, and launched it into self-sustained growth.

But to show how geology came to have a form and dynamic of its own is not to divorce it from the wider social context, which was precisely the condition of its existence as an organized science. Geology being a fabric which was made, we must understand its development in terms of the society which made it, or at least permitted its making. The Earth has been conceptualized differently in Europe from in China. So, equally, geology developed differently in Britain from in France, Germany, Italy or Scandinavia; differently in England from in Scotland or Ireland (cf. Merz, 1896–1914; Crosland, 1975).

Three final points may clarify my approach. Firstly, my subject is the transition from earlier beliefs about the Earth and ways of investigating it, to the science of geology as practised in the nineteenth century, a practice with many affinities to current geology, which produced much knowledge the truth of which nobody disputes. Salutary broadsides have lately been fired against Whiggish, teleological, history of science (cf. Thackray, 1970b). These timely warnings must serve as a *terminus a quo* for future history. But if historians take them too doggedly to heart, history of science will fastidiously reduce itself to the opposite vice, of timidly tackling each scientist, idea or paradigm in splendid isolation, and will lose its sense of the historically significant. Naturally the historian must initially understand men and ideas on their own terms (Skinner, 1969). But then he must also evaluate their function in larger historical movements which transcend the individual. The history of so-called cul-de-sacs should certainly be explored – Biblical science, the fate of alchemical mineralogy, or 'mere' fossil collecting. When it is, many may be found to be highways after all. But the central feature, hence the central historical problem, of modern science has been its power unceasingly to expand – technically, intellectually, ideologically and socially. The historian must grasp the nettle, and probe the *historical* causes for the power of the scientific movement. To investigate the progress of science is not to hymn it, but it is to acknowledge the historian's urgent imperative to study the 'progressive' aspects of human life (Carr, 1961; Plumb, 1964).

Secondly, it is not the point of this book – nor is there room – to biographize individual geologists or to expound their discoveries and theories. The aim is rather to present an inevitably programmatic interpretation of how the science of geology was made. The subject matter is necessarily dynamic, and elusive – though, I hope, not illusory. The argument cannot be proven by secure weight of documentation as in chronicling a controversy or reconstructing a life-history. In its nature, my case rests upon the general plausibility of the interpretative pattern for a lengthy time period. Though, like a moving picture, composed of many stills, it is to be judged by the animation. For the scenario to cohere, it must cover a century and a half, which inevitably means that many individuals and their work, and innumerable problems, have only walk-on parts. I hope partly to fill these gaps by publishing in the near future a number of articles covering in detail topics barely touched on here.

Lastly, what are the broader goals of this study? It is meant as an illustration of how a science is created, developed and maintained; how urgent natural problems change over time; and, perhaps most important of all, how complex a fabric is the activity of scientific investigation, although, iceberg-like, only the tip ever appears on the public surface. It is offered as a contribution to grasping some central problems in the relationship of science to Western culture, and the emergence of modern patterns of thought. Why did natural science expand so remorselessly, as a vocation and belief system, throughout this period? How should we interpret its powerful, equivocal role in the forming of modern civilization? I hope this study of the constituting of the Earth as an object of scientific study throws some light on these issues.

I should be fortunate if this work helped to persuade geologists – and scientists in general – that the history of science is much more than a tale of anticipations and precursors, founding fathers and who discovered what first. I should be delighted if it helped historians in general to see that the history of science is continuous with their own economic, social, political and intellectual history, though with certain distinctive historical features. Not least, of course, I hope it might offer historians, sociologists and philosophers of science some food for thought.

# I

# Orientations: *c.* 1660–1710

## *The Renaissance geocosm*

Beliefs about the Earth – its history, form, purposes and future –
have always been central to European culture, religion and science.
Discourse on the Earth was common in Renaissance Britain, de-
riving largely from Classical sources, passed down by Medieval
scholars, and ranging from concrete fact to cosmological wisdom.
Proper ideas about the Earth mattered in a largely rural society,
whose ideology was grounded upon cosmic order, whose understanding of man hinged upon macrocosm–microcosm analogies,
and whose sacred book began with its Genesis (Humboldt, 1845–
62; Glacken, 1967). Such common-sense knowledge was socially
broadcast in sermons, school-books, compendia and folklore, re-
verberating through imagery, symbolism and metaphor (Duncan,
1954; V. I. Harris, 1966; Williamson, 1935). Much of it carried
over into the latter part of the seventeenth century, being debated, modified, accepted or rejected, and serving as the fruitful
nub of future problems.

But in light of later developments, certain features stand out.
Little empirical investigation of the Earth was conducted in
Britain up to the mid-seventeenth century (Raven, 1947: 328).
Most data derived from Continental sources. Knowledge of the
Earth was generally located within other bodies of discourse (e.g.
meteorology, or Biblical exegesis) rather than being organized as
a science in its own right. What empirical study there was centred
upon particular products, such as gems, rather than upon the
Earth itself. Furthermore the diverse traditions of discourse about
the Earth now to be discussed were conducted largely in mutual
isolation.

Before the mid-seventeenth century, discussion of the Earth as a physical body primarily regarded it as a planet within general natural philosophical cosmology and cosmogony, Biblical chronology and natural theology. There was no independent science of the Earth as such. Of course, the most prominent ideas within these traditions – such as the cycle of Nature, the Earth's decay, and deluges – form an audible bass-line in the scores of later geologists, as indeed did many of the early *problems*, such as the Earth's age, its economy, its adaptation to man. Cosmogony and Scripture formed a matrix of later theories of the Earth, one which – despite a once-influential historiography of the 'conflict' of Christianity and science – proved geologically fruitful (Tuveson, 1949). However, though subsequent theories grew out of earlier Biblical and general natural philosophical inquiry, the distinctions between the traditions are real. Thus, in early Stuart England, the history of the sublunary sphere was investigated within the cosmogonical problem of the origin of the entire universe, of Creation itself, and its future within a cosmic chiliasm. The question of the Earth was subsidiary. The Earth was viewed within astronomy, astrology and meteorology. The *element* Earth, or the *planet* Earth, rather than its crust, occupied natural philosophy (cf. Kargon, 1966). Not till the total triumph of heliocentrism could the Earth's status begin to be detached from that of the whole universe. Only the age of reconnaissance and the first circumnavigations created some familiarity with the geography of the entire globe and directed interest towards the total surface of the Earth (Grierson, in Blacker and Loewe, 1975).

Furthermore, early struggles between rival cosmogonies adduced chiefly philosophical evidence – the relative *a priori* plausibilities of Aristotelianism, Platonism, Hermeticism, atomism and, later, Cartesianism. Or, the weapons were philological and exegetical, depending upon the canons of interpreting Biblical, Patristic and Classical texts (Meyer, 1951; A. Williams, 1948). Problems and solutions sprang from scholarly inquiry into the written word, rather than examination of the Earth itself: 'Mineralists look exactly into the twenty eighth of Job' (Sir T. Browne, 1928–31, v: 4). Scholarly *explications de texte* and rational philosophy to prove Creation continued, of course, through into subsequent centuries. The traditions of Edward Stillingfleet's *Origines sacrae*

(1662) and Matthew Hale's *The primitive origination of mankind* (1677) were later sustained by Hutchinsonianism and 'scriptural geology' (A. J. Kuhn, 1961; Millhauser, 1954). The point is, however, that rival forms of discourse developed alongside these, grounded upon meticulous examination of the Earth, and pursued by men who chafed at older traditions. John Ray significantly could 'never find in my heart to read' Hale's book, 'not expecting any great matter' (Gunther, 1928: 270).

Another long-established mode of inquiry into the Earth was the geographical, topographical and chorographical, expounded in the work of such travellers and scholars as Lambarde (1576), Leland (1745), Camden (1586) and Carew (1602). But up to the Civil War, this tradition had little probed the Earth's structure. Topographers generally celebrated objects human rather than natural – antiquities, civil and family history, houses and heraldry. Even a pioneering work such as Gerald Boate's *Ireland's naturall history* (1652) focused primarily on physical relief – coast-lines, rivers, harbours and mountains – and on flora and fauna, rather than rocks (C. Webster, 1975: 428f.). Only rarely, as in Carew's account of Cornish tin mines, were the composition and structure of the Earth important (1602: 8–27). Disputes over the interpretation of terrestrial features frequently hinged upon scholarly authorities rather than personal investigation (E. G. R. Taylor, 1934; Challinor, 1953).

No wonder, therefore, that the Earth's interior retained its secrets, given the prevailing ignorance of its surface. Geographers had hardly plotted the physical features of much of the kingdom. Nor could they satisfactorily represent relief on maps.[1] Topographers no doubt provided a matrix for the growth of geology, yet many traits of their work were later to be rejected. They singled out the unusual rather than the norm, and particularly marvels drenched in human associations. They passed on, often uncritically, folklore explanations. Local curiosities were favourite objects of interest – Bristol 'diamonds', Whitby ammonites, holy wells – rather than the Earth's crust itself (Childrey, 1661; S. Smith, 1943).[2]

Parallel in many of its intellectual traits was English natural history. Enumeration of the Earth was neglected by such leading naturalists as William Turner, John Gerard and Thomas Johnson, their passion being the animal and vegetable kingdoms (Raven, 1947; Hoeniger, 1969). General naturalists' service to later Earth science lay in transmitting and popularizing Classical and Renaissance knowledge in their books of natural history, wonders and secrets, lapidaries, pharmacopoeias and encyclopaedias. In content and form, their beliefs about the Earth typified Renaissance assumptions – whether Aristotelian, neo-Platonic, Hermetic or eclectic – about the natural order's array of correspondences, emblems, macrocosm and microcosm analogues, animism, magic, teleology and anthropocentrism. In this way of seeing the world, the nature of a fossil or mineral object, absolute in itself, its purposes and powers dictated by Creation, and its symbolic significance, all meant more than its historical and physical relation to adjacent objects (Foucault, 1970: ch. ii). This ontological perspective held out to later generations certain ideas which seemed fruitful (e.g. that crystals grew), and others which seemed retrograde (e.g. that fossils were *lapides sui generis*). But it was a philosophy of Nature which was increasingly discounted as erroneous and irrelevant by successive generations of Bacon-faced English naturalists. Attitudes were not of course revolutionized overnight. Sir Hans Sloane's mineral collections bulged with objects whose meaning lay in their magical and human associations (Sweet, 1935). But there is a marked and continuing shift between what, say, the popular Tudor *Book of secrets* (1550), Nicols's *Lapidary* (1652), Plot's *Oxford-shire* (1677) and Woodward's *Fossils of all kinds* (1728) recognized as important knowledge about rocks and minerals. Simultaneously, strands of Protestantism as well were rejecting 'superstitious' animistic cosmologies which saw the Earth as enchanted or haunted and which recognized 'holy places', in favour of utilitarian and natural law perspectives (Thomas, 1971: 41, 70, 274).

Certain characteristics genuinely distinguish these from later natural historical perceptions of the Earth. Earlier works were more concerned with mineral objects in themselves (or as linked to universals), than with specific geographical–geological location. Thus, encyclopaedias tended to carry distinct entries on particular minerals and gems, such as cornelian (e.g. Nicols, 1652: 123f.), rather than enumerating the minerals found in a particular area. This is scarcely surprising, in view of the international sources of information ransacked for encyclopaedic compilations. But it does mean that Renaissance British natural history had little ambition to relate mineral products to their local matrices or to other minerals. No important classification of the mineral kingdom was constructed in Britain before mid-century (St Clair, 1966: ch. i). Likewise, the magical properties, the anthropomorphic associations and medical powers of fossils were given more attention than their physical nature (Oakley, 1965; Nelson, 1968). Lastly, the information contained in English natural histories had often been pillaged uncritically from Classical, Medieval and Continental Renaissance texts. Even where not originally inaccurate, errors of transcription, translation and scholarship had often infiltrated. Britain produced no Renaissance scholar–naturalists for the mineral kingdom of the calibre of Gesner, Agricola, Palissy, Kentmann or Aldrovandi.

Alchemy and chemistry promised an important impulse for investigating the Earth (Debus, 1965; 1977). Their commitment to empiricism and experimentation prefigured later methods. Alchemy's ideology pledged co-operation between chemists and miners, and promoted experimental traditions in closely related fields – e.g. the medicinal analysis of mineral waters (Debus, 1969). Gabriel Plattes in particular used laboratory experiments to replicate formation of minerals and strata in Nature (Plattes, 1639; Debus, 1961). Such work could be undertaken partly because alchemy had confidence in the micro–macrocosmic parallel between transformation of chemical agents in laboratory experiments and similar active processes continuously operating in the great laboratory of the Earth, whose supposed central heat was analogous to the adept's furnace. Alchemical traditions sharply underlined the Earth's active powers, and thus fostered interest in fire and water, and dynamic processes such as the growth and change of state of minerals, earthquakes and volcanoes. At the same time, Paracelsianism was translating Genesis into alchemical language, and so establishing an important tradition of chemical speculation on the formation of the Earth (Debus, 1966; 1974).

Paracelsian alchemy might have spawned a flourishing school of Earth investigation. Indeed, certain types of analysis (e.g. of mineral waters), and particular later theories (such as Thomas Robinson's *Anatomy of the earth* (1694) and William Hobbs's

'The earth generated and anatomized' (1715; Porter, 1976b)), were heavily indebted to it. But, as Debus has argued, comprehensive loyalty to Paracelsianism never took root in Britain, and in late Stuart times it came under mounting fire from the mechanical philosophy, with the result, for example, that Robinson's theory was either ignored or ridiculed (North, 1934). Hence, except in mineralogy and metallurgy Renaissance chemical philosophy did not give rise to dominant lasting British traditions of inquiry into the Earth.

Practical mining might in a similar way be *prima facie* expected to have opened up research. Elizabethan England saw expanding mineral and metal exploitation under the leadership of Germans, trained in Agricolan skills. While this in itself betrays a lack of sophistication amongst native miners, it could have expedited transmission of superior foreign skills to form a permanent British technical expertise (Collingwood, 1912; Donald, 1955; 1961). Largely for financial reasons, however, mines enterprises declined in the seventeenth century. German miners returned home, leaving a vacuum of skill (Gough, 1930). Successful native mines venturers were men of business acumen rather than expert knowledge of the Earth, and they produced little mining literature.[3] Traditional miner's earthlore was little tapped or refined by elite natural philosophers.

Early Stuart Britain, then, presents a culture and a body of natural knowledge in which beliefs about the Earth were prominent. The Earth was charged with moral, religious and human significance. Stones contained sermons. Scientific ideas passed down from Classical and Medieval times offered both comprehensive internally coherent general philosophies of Nature and also explanation of particular phenomena such as earthquakes, volcanoes, fossils and the production of minerals and landforms. Yet little organized empirical investigation into the Earth was being undertaken. Certain endemic social conditions such as difficulties of travel (Parkes, 1925: ch. vii) and the disruptions of the Civil War clearly did not encourage study. Neither did prevalent attitudes, such as that to study the Earth was demeaning (J. Webster, 1671: Preface), or widespread indifference to landscape and external Nature (Nicolson, 1959: chs. i, ii). In the universities – as at Gresham College – no teaching was provided expressly about the Earth (P. Allen, 1949; Adamson, 1976; Frank, 1973). Long after botanical and medical collections developed, mineral collections were lacking. No eminent British man of science had yet specialized in Earth science.[4] Bacon kept unaccustomedly quiet in this field. Charles Webster's persuasive argument (1975) for the general flourishing of natural philosophy notwithstanding, investigation of the Earth was one area of natural knowledge which scarcely advanced in Britain in the first half of the seventeenth century.

## Institutions and ideals

Investigation of the Earth did not prosper before effective communities and institutions materialized. In Continental Europe, princes maintained scientific institutions to serve their economic, technical, bureaucratic and military purposes (cf. Ornstein, 1913; Hahn, 1971). In German principalities, in particular, they went hand-in-hand with mining. In Britain, the private enterprise of natural philosophers in founding the Royal Society of London in 1660 first put natural investigation on a formal and stable institutional footing. From that time, study of the Earth began to attain clearer direction and coherence. Even without a Royal Society, *ad hoc* groups interested in exploring the Earth would probably have continued the impulse given by the upswing of activity during the Interregnum (C. Webster, 1975). But the existence of the Society gave these stature and stability and, not least, a unique outlet for publication. Admittedly, it did not act as a corporate director of research into the Earth, being more effective as an entrepôt of knowledge (cf. Stearns, 1970; Porter, 1972). Its function was to orchestrate thorough empirical research rather than to act as midwife to geology.

The Royal Society promoted investigation chiefly in two respects. First, it served to *initiate* study and collect information. To this end, its early ideology, a Baconianism expressed in its formal documents, in Sprat's *History* (1667), and in action by such members as Oldenburg, Hooke and Boyle, was particularly apposite. It stressed the capacity of everybody to collect and communicate useful information; the importance of co-operation – between metropolis and province, fact-gatherer and theorist, amateur and professional; the national and utilitarian value of empirical natural history; the indispensability of accurate, critical, first-hand data (Jones, 1936; M. B. Hall, 1975; Stearns, 1970). Such desiderata successfully galvanized collection, preservation and publication of basic facts. Even elementary knowledge of the Earth had been so parlously deficient that the slightest papers could nevertheless be much used.[5] Widespread amateur involvement enabled the Royal Society to play a more substantial part in the development of Earth science in Britain than did the Académie Royale des Sciences in France.

Secondly, the Society's activities facilitated constructive use, evaluation and publicizing of the facts communicated. Particularly important were debates at the Society's regular meetings. Topics reported and discussed in the first generation of the Society include coal mines and coal damps; microscopic examination of fossils and other mineral substances; divining rods; subterraneous trees; the growth of minerals; the location of mines; earthquakes; new Continental publications, such as Steno's *De solido*; the origin of metals; subterraneous fire; different building stones and their uses; the identification of fossils; the need for mineral maps, and Hooke's theories of the Earth – together with profuse reports of mineral phenomena and substances, and repeated presentation of specimens and miscellaneous experiments on mineral products (Birch, 1756–7).

Next, the Society built up the first substantial, accessible public collection of natural specimens in England. Its reputation became impressive – being called by Andrew Balfour a 'very fine collection of natural rarities' (Murray, 1904, i: 131) – and Nehemiah Grew's published catalogue (1681) served the future as an important index of specimens. The founding of the Society indeed underlined natural history's dependence upon collections: 'It were. . . much to be wishht [*sic*] for and endeavoured', wrote Robert Hooke, the Society's demonstrator and curator,

that there might be made and kept in some Repository as full and compleat a Collection of all varieties of Natural Bodies as could be obtain'd. . . The use of such a Collection is not for Divertisement, and Wonder, and Gazing, as 'tis for the most part thought and esteemed. . .but for the most serious and diligent study of the most able Proficient in Natural Philosophy. (Hooke, 1705: 338)

Henry Oldenburg's and Robert Boyle's personal contributions were vital. Oldenburg's *Philosophical transactions*, a personal venture, were of capital importance. For Earth science, more than classical botany or zoology, is a study of irreducibly diverse local phenomena. These, as facts and as problems, were often first brought to general attention by notices in the *Philosophical transactions*. But Oldenburg's journal did not merely pioneer accumulation of local knowledge; it encouraged its transmission in the novel, concise form of the scientific article, in plain, neutral prose with diagrams, figures, tables and measurements (Kronick, 1962). Under Oldenburg's editorship, the *Philosophical transactions* publicized foreign works such as Athanasius Kircher's *Mundus subterraneus* (1664–5) and Steno's *De solido* (1669) – which then, by provoking Martin Lister, sparked off major controversy in England on the origin of fossils (Lister, 1671). The journal also brought to the public eye for the first time many striking British phenomena – such as the Giant's Causeway and fossil plants ('lithophytes').

Possibly Oldenburg's and Boyle's major service was in producing a succession of 'Queries' on the natural history of the Earth, chiefly the 'General Heads for a Natural History of a Countrey' (*Phil. trans.*, 1666a), the 'Articles of Inquiries touching Mines' (*Phil. trans.*, 1666c), and the 'Directions for Inquiries concerning Stones and other Materials for the Use of Buildings' (*Phil. trans.*, 1673). Responses also appeared (Glanvill, 1667; 1668). Furthermore, they triggered off the use of questionnaires amongst naturalists such as Plot (who in his two county histories explicitly followed Boyle's scheme), Woodward and Lhwyd.

The *Philosophical transactions* Queries are learned and pertinent. Admittedly, their assumptions still embody Renaissance cosmology (e.g. concern whether there are 'subterraneous daemons' (1666c: 343)). And, compared with later questionnaires, they seem a random accumulation rather than an inquiry directed to a particular end. But their real significance lay in their stress on first-hand, quantified observations.

In addition, Oldenburg personally sent many Queries to correspondents – especially to those living or going abroad – with a view to publishing findings in the *Philosophical transactions*. 'I could wish you had with you those Heads of Inquiries yt are printed in severall Numbers of the Phil. Transactions', Oldenburg wrote to John Downes in 1669 (Hall and Hall, 1965–, v: 315).[6] Also, as secretary, he was able to solicit contributions to the journal. Some were from men who probably would never have published had there been no Royal Society – such as the fossil-collecting Devonshire gentleman, Samuel Colepresse (Hall and Hall, 1965–, iii: 457; Colepresse, 1667; 1671); others from those whose work would otherwise have been less easily available to an English audience – Borelli (1671), Septalius (1667), Scilla (1696).

Arguably, the founding of the Royal Society stimulated Earth science more than most, since it had lain in so rudimentary a state, and – by its nature – needed a large body of devotees over a wide terrain. The Society's role however was not to hatch theory, but rather to boost the empirical investigation of Britain.

Other societies followed which provided a forum for discussion of the Earth. Natural history, and especially minerals and fossils, were central to the activities of the Oxford Philosophical Society, founded in 1683, because of the energetic husbandry of Robert Plot, keeper of the newly established Ashmolean Museum (Gunther, 1925a; 1939). Plot being independently one of the Royal Society's secretaries, papers readily flowed between the two institutions – harbinger of many links between metropolis and provinces. The Oxford Society's activities paralleled the Royal Society's, with discussion of papers and specimens, received from resident and non-resident members. Earths, mineral waters and petrifactions peppered the debates, partly because of Plot's attentions, but also because they formed ready objects of interest both for provincial correspondents like Andrew Paschall and John Beaumont, and for virtuoso dons who graced the meetings. The society co-ordinated the work of far-flung individuals, re-introduced organized science to Oxford, and channelled specimens towards the Ashmolean. Yet, Plot's organizing zeal as 'prime-mover' removed, the society soon collapsed. For at this stage the mere establishment of useful institutions was no guarantee of permanence.

Its twin was the Dublin Philosophical Society (Hoppen, 1970). Like the Oxford Society with which it corresponded, it took its cue from the Royal Society, soliciting publication in the *Philosophical transactions*. It, too, was the brainchild of individuals –

William and Thomas Molyneux – and died with them. It sought
to unite the social elite of Ireland as 'common scientists' in quest of
natural knowledge, and to disseminate it for utilitarian, political
and cultural ends.

The society's goal was a Baconian natural history of Ireland,
including a general survey, a knowledge of mineral resources, and
publicizing of the country's unique wonders. But the ambition
outstripped achievement. Instead of a comprehensive survey, the
society's legacy was a mere handful of stimulating papers on the
peculiarly problematic features of Irish topography – the petrifying waters of Lough Neagh, the gigantic mineralized bones of
the Irish elk, and of course the Giant's Causeway. The Dublin
Society (f. 1731) later picked up its unfinished work, but no
thorough geological survey made progress in Ireland before the
work of C. L. Giesecke and Richard Griffith into the nineteenth
century (J. M. Eyles, 1961; Berry, 1915).[7]

Other local scientific societies were mooted. One which was
formed was the Somerset Philosophical Society, organized by
Andrew Paschall.[8] His aims encompassed familiar Baconian
aspirations for local natural history. Had the society prospered,
the mineral kingdom would, no doubt, have been prominent, for
the mines of Mendip were close and John Beaumont and Joseph
Glanvill were members. But the point is that it dwindled. In the
late seventeenth century scientific societies could only survive in
the metropolis and, fleetingly, in university and capital cities
('Espinasse, 1958). Study of the Earth could advance, but the
future was most precarious.

Outside new societies, other institutions could indirectly support
study of the Earth – such as Gresham College which provided a
physic chair for John Woodward. Despite the popular stereotype,
Oxford and Cambridge housed growing numbers who had strong,
though generally secondary, interests in the mineral kingdom
(Frank, 1973). In Oxford, Robert Plot, Edward Lhwyd, Joseph
Archer and Arthur Charlett stand out; in Cambridge John Ray,
and later John Covell and William Vernon (Gunther, 1925;
1937). They made observations, built collections, corresponded
with others, and absorbed contemporary literature (Gunther,
1925: 317, 337). In Oxford, the founding of the Ashmolean
ensured some continuity, as did the Woodwardian chair in Cam-

bridge after 1729. Except alchemically, Elias Ashmole had not particularly cultivated minerals, and his collection, acquired from the Tradescants, had been initially botanical (M. Allen, 1964). But the fact that the first two keepers of the Ashmolean, Plot and Lhwyd, specialized in the mineral kingdom testifies to the rise of mineral investigation. In default of an Oxford chair, the keepership and the collection helped to secure future study (Jahn, 1962–8).[9] Oxford University as such, however, gave no encouragement. It declined to advance money to purchase collections – such as William Cole's – for the Ashmolean, or to publish Lhwyd's *Lithophylacii.* Lhwyd received no university salary. Neither English university made any concession in its undergraduate courses to Earth science.

Without foci like the Ashmolean or the Oxford Philosophical Society, Earth science in other universities hinged upon individual enthusiasts. Andrew Balfour and Robert Sibbald gathered a collection in Edinburgh, towards a grand project for a Scottish natural history (Sibbald, 1684; 1697). Robert Wodrow in his student, virtuoso phase in Glasgow avidly collected fossils (Sharp, 1937). Beyond universities and the societies discussed, no other *formal* institutions existed at this stage to foster study of the Earth.

Nevertheless, one other semi-permanent medium was emerging to provide some continuity: the collection and the private museum. Before the mid-seventeenth century, Britain had none of those prodigious natural history collections which graced Renaissance Italy, Switzerland and Germany, most British collections having been of human artefacts and *objets d'art* (Wittlin, 1949; Bedini, 1965). Later Stuart England saw the rise of the general private virtuoso collection, which would include natural history objects, such as John Evelyn's or Ralph Thoresby's; and in particular the appearance of ordered, catalogued, didactic natural history collections, like James Petiver's, William Courteen's, Hans Sloane's or Richard Mead's. The logic of specialization led to mineral and fossil collections, such as the much admired ones of William Cole, John Beaumont and John Woodward.[10]

This development marks the growing autonomy of Earth science in Britain. Woodward's collection provided the basis for the most complete mineral classification yet attempted (1728; V. A. Eyles, 1971). During their founders' lifetimes, many were open to public inspection, and on their deaths, most were preserved. Much of Woodward's collection went to Cambridge, where, augmented through the eighteenth century, it still survives. Sir Hans Sloane grafted several of them onto his own, which, bequeathed in 1753, then became the trunk of the British Museum. The growth of mineral and fossil collections provided a durable physical basis for future inquiries (cf. Sweet, 1935; de Beer, 1953; Jahn, 1962–8).

### *Men and society: invisible colleges*

But who were the investigators? What was the nature of their community, and of their relation to society at large? The *causal* question, why certain people – and not others – began to give some impetus to investigating the Earth, is obviously fundamental to the inquiry. We must, however, initially rest content with descriptive generalizations. Earth science was the beneficiary of the increasing popularity of science across the board. There are no unique reasons why men investigated the Earth, other than the broad reasons for which they studied Nature. This is so precisely because the Earth was investigated at this time within the discipline of comprehensive philosophy. Here is not the place to rehearse the numerous – partially true – explanations for the increased tempo of natural inquiry. It is more profitable to establish which sorts of men were now investigating the Earth, and how the social character of that fraternity shaped the form of science they professed and performed.

The prime social determinant is that the great majority of important investigators were university educated and/or medically trained.[11] This does not mean that universities positively encouraged science. Rather, it strongly suggests that other kinds of background and training were unpropitious environments for scientific inquiry into the Earth. Almost no one outside university and medically trained circles was publishing any major contribution to Earth science. There were no groups composed chiefly of, or led by, non-university and non-medical men. Former students who settled in the provinces did their scientific work in correspondence with other *alumni* (in several cases, former college colleagues) living distantly. They scarcely worked within indigenous local groups of devotees – for these did not yet exist. 'I wish I could any way give you assistance by procuring promoters', Ray lamented to Lhwyd, 'But truly there are not in our neighbour-hood any of ye Gentry or Clergy yt minde any things of Natural History or ingenious Literature' (Gunther, 1928: 220). Practical involvement with the Earth – farming, prospecting, mining, sur-veying – formed no springboard for scientific inquiry. At this stage men from non-liberal backgrounds and professions were thought to lack the necessary habit of curiosity. 'The truth is', wrote Burnet,

the generality of people have not sence and curiosity enough to raise a question concerning these things [mountains] or concerning the Original of them. You may tell them that Mountains grow out of the Earth like Fuzz-balls, or that there are Monsters under ground that throw up Moun-tains as Moles do Molehills; they will scarce raise one objection against your doctrine. (Burnet, 1684: 140)

Many major investigators were not merely university educated, but were Fellows of colleges or professors, for part or all their careers.[12] They formed tightly-knit groups, united not by eco-nomic and social *origins*,[13] but by shared educational experience. Excepting the Scots, they were Anglicans almost to a man, many in Orders. Apart from Ralph Thoresby, none was a Dissenter.[14] Thus while study of the Earth was a minority activity, it was not the pursuit of 'outsiders'. Granted, several were on the liberal and rationalist wing of the Anglican Church (most notably Burnet and Whiston), but this was not abnormal within the contemporary intelligentsia. The great majority had received their training in the orthodox, establishment-approved vocations of medicine or divinity.

This shared liberal educational background also fostered com-mon habits of thought. University humanism inculcated respect for scholarly forms of scientific discourse and for the authority of the written word. They envisaged Nature as the book of God's works. Their earliest ideas about Nature, Earth history and fossils came not from fieldwork, however committed as fieldworkers they later became. Their training also led them to study the Earth in the intellectual context of a panoply of other natural and human-istic disciplines – medicine, natural theology, antiquities, chrono-logy and comparative mythology.[15] There was little specialization at this stage. Neither did they earn a living from their knowledge of the Earth. Lhwyd thought natural history's snail-like progress was because few could devote their whole time to it (Gunther, 1945: 364). Several abandoned or pared down their investigations under professional pressure: Wodrow in the Church, Lister in medicine.

Not surprisingly, this compact university elite pursued Earth science using the range of styles and techniques already established within the scientific tradition. One category – for example Burnet, Whiston, Halley and Keill – was seeking to establish the theory of the Earth as an exact science within natural philosophy with, some hoped, mathematical precision. Peering at stones for them was confused and ambivalent in value, and they had little contact with the fieldworkers of the age. They hoped to derive a theory from fundamental physical, cosmological or astronomical postulates. The same is true, *mutatis mutandis*, for chiefly religious theorists like Warren and Croft. Most theorists were not fieldworkers, and most fieldworkers did not write theories of the Earth.

Secondly, there were those like Ray, Beaumont and Sloane who were certainly competent fieldworkers, but whose essential work lay in patient reconstruction, identification and classification of specimens indoors, the specimens themselves often being provided by others. These were among the leading taxonomic natural historians of the age. They studied the Earth through specimens much as they studied plants and animals. They did not pioneer special approaches to the Earth.[16]

Thirdly, a growing number of topographers, antiquarians, local natural historians, travellers and collectors – scarcely present before mid-century – were responsive to the ideals of the new Baconian science. They travelled (however locally), observed at first hand, and compiled materials for regional natural histories. This important new body of observers includes Abraham de la Pryme, John Morton, William Nicolson, Robert Plot, Charles Leigh, Richard Richardson and Martin Lister (Carr, 1974). None of them revolutionized the study of the Earth, and collectively their importance lay less in what they achieved in their lifetime than in providing materials for the future. But they were beginning to expose the problems and pioneer the methods which were to become characteristically geological.

Lastly, two men in this period stand apart in style and methods: Woodward and Lhwyd. Each in his own way – and they came to dislike each other heartily – was profoundly dissatisfied with contemporary Earth science, and tried single-handedly to transform it in directions which were manifestly suggestive to later geologists. Both were passionately committed – though not to the Earth alone, for Woodward was equally devoted to antiquities, and Lhwyd to his Celtic studies, the *Archaeologia*: one must not by prolepsis make them into specialists. Lhwyd sent himself to an early grave with work and travel. Woodward's pride, jealousy and independence; the fastidiousness with which he kept his collection;[17] the stringent rules he laid down for the Woodwardian chair's incumbent, to ensure *his* dedication to work (Clark and Hughes, 1890: ch. v); and not least his prolific output and enormous collection – all betoken furiously engaged and directed energies. Both avoided social pleasantries. Neither married.

Lhwyd's vision, which pinpointed the Achilles heel of contemporary investigation, was that current knowledge was speculative, and based upon far too narrow an induction. He was the first to emphasize that the problems posed by the labyrinthine complexity of the Earth could only be solved by exhaustive travel (Gunther, 1945: 13, 174, 269–71). While Plot stuck to Oxfordshire and Staffordshire, Leigh to the Peak, and Morton to Northamptonshire, Lhwyd was exploring Wales, Cornwall, Devon, Brittany, Ireland, the Isle of Man, the Lake District, Scotland. It goes without saying that Lhwyd had scant interest in many of the problems central to later geology – the strata, the history of life, landforms. But he was the first to develop the *idée fixe* of travel, which became cardinal to the later science (Emery, 1971).

Woodward was remarkable in his day – and prophetic – for breaking down boundaries between methods and disciplines. He was an assiduous field-observer, yet presented his observations within a Newtonian theory of the Earth. He studied the strata, yet also constructed classifications of minerals and metals. He undertook minute analysis of fossils, but also offered a comprehensive account of their organic nature and history. Woodward's theories were ridiculed and he was personally vilified. But the lonely furrow he ploughed left a signal mark for later work (Ward, 1740; Jahn, 1972; V. A. Eyles, 1971).

What kinds of subcommunities and infrastructures did these investigators create for themselves? The Royal Society, as already argued, provided one institutional focus for the science, serving as map and compass for its members. But the Society did not actively establish contact between worker and worker, particularly those in the regions. The Society was metropolitan, but its Fellows far-flung. Most craved support, advice and suggestions from fellow enthusiasts (cf. R. Richardson, 1835: 70f.). Their correspondence rings out with pleas for more communication; for personal introductions, for exchange of ideas, books, specimens, information. Most felt isolated and sought community. How were the patterns of contact forged significant in shaping the deeper development of the science? Will they help to explain the kind, and quality, of the work done?

Most of those who wrote *theories* – or counter-theories – of the Earth, such as Burnet, Whiston, Warren, Keill and Arbuthnot, had close contact neither with fieldworking naturalists nor co-operative relations with each other. The 'theory of the Earth' was an individualist and self-contained form of expression, within the genre of controversial literature in a deeply polemical age (Holmes and Speck, 1967). Woodward bridged that divide. Though a theorist and an arrant individualist, he played entrepreneur to a number of fieldworkers. Many of these he treated as inferiors, fit only to feed him information on demand, and confirm opinions on pain of not contradicting the master (Woodward, MS Gough Wales 8). On the other hand, his charisma won him converts such as John Morton and J. J. Scheuchzer, admirers like William Nicolson (Lancaster, 1912: 97), and disciples such as John Harris, Benjamin Holloway and, initially John Hutchinson. Woodward also paid servants to work for him (Lancaster, 1912: 115). His monomania was both milestone and millstone for eighteenth-century study of the Earth. The authoritative clarity of his ideas offered a heuristic framework for many observers such as Benjamin Holloway and Alexander Catcott. Yet his brooking no criticisn prolonged a situation in which high-level theories and Baconian natural history ran parallel courses with only tenuous relationships between each other.

More auspicious were the circles of natural historians who corresponded on terms of equality and friendship. An interlocking network of correspondence was built up between John Ray, Martin Lister, Robert Plot, Hans Sloane, Tancred Robinson, William Nicolson and Edward Lhwyd, with men like William Vernon, Jabez Cay, Samuel Dale, Joseph Archer and John Morton somewhat more peripheral (cf. D. E. Allen, 1976: ch. i). Each corresponded with several others directly, and additionally picked up news of others' activities through mutual friends. Through their letters they sought advice and criticism, discussed their findings, requested identification of specimens and exchanged books and specimens.[18] Before indigenous local scientific communities arose, and where travel was too difficult or expensive, ties of correspondence were for many the sole scientific contact. It is a poignant thought that Lhwyd and Ray never met. Yet the work of both was enriched by lengthy exchange of letters. The stimulus of such contacts gave the work of this circle critical sharpness, and a sense of direction and purpose.

Contact of friends and equals by letter reinforced Baconian, empirical natural history. The letter, like the scientific paper promoted by the *Philosophical transactions*, was the ideal medium for conveying information and posing queries, but it discouraged ambitious theories. Ties of friendship encouraged cautious and courteous expression of ideas: accommodation not assertion. The letters in particular of Ray, Lhwyd and Lister show open-mindedness, an ostentatious (and occasionally self-righteous) distancing from theorizing, and an admission of the refractory problems confronting an empirical science of the Earth. Community by letter broadened the intellectual vistas of those whose geographical horizons were drastically limited, and created a realistic awareness of the multifariousness of the Earth.[19]

On the other hand, reliance on correspondence betokens a halting community. Even ignoring the grave inconvenience of lost letters and parcels, many naturalists were normally working without daily face-to-face contact. In particular, almost all fieldwork was solitary.[20] More seriously for the long-term, the absence of personal contacts with *local* communities of devotees meant that these naturalists failed to recruit and train successors. There were no pupils of Ray, Lister, Morton, Lhwyd. The community by correspondence was too personal an association to outlive its members.

Lastly, what were the relations between society at large and the increasing numbers who were studying the Earth? Many trends in British social and intellectual life were stimulating the pursuit and use of natural knowledge. The growth of commercial capitalism could promote the advance of science, and looked kindly on it for utilitarian reasons. The Civil War's discrediting of religious fanaticism and dogmatic theology passed the normative mantle to natural knowledge. The climate of the Ancients versus Moderns debate registered growing confidence in man's capacity to advance in knowledge not through deference to past authority but through his own endeavours. All such stimuli kindled the general range of natural philosophy of which Earth science was an integral and dependent part (Redwood, 1976: chs. iv, v).

In this period, however, it would be a glib and trivial over-simplification to speak as though 'Society' 'required', or powerfully 'stimulated', Earth science. This is not to imply a – thoroughly anachronistic – 'conflict' relationship between the supposedly modern, enlightened scientist and a supposedly benighted, theology-and-classics ridden society. Public reception of the science drew limited hostility from certain quarters – theological, humanistic and social.[21] There was, however, no mortal combat between Earth science and theology or letters, but a debate over proper demarcations in an age when demarcations were particularly insecure. Theories of the Earth aroused fashionable interest rather than torrents of public consternation.

Nevertheless, from the longest historical perspective, *indifference* probably best sums up the social reception and standing of the science at this time. Earth science was small beer when compared with struggles between Stuart and Hanover, Whig and Tory, Anglicans, Catholics and Dissenters. In so far as it was *part* of such debates, its part was small (though see Jacob, 1976). Very few investigators received public recognition, position or payment. They laboured long to fill up their subscription lists. Lhwyd complained that the Welsh gentry would readily subscribe for antiquarian but not for natural history books. He had acute difficulty publishing his *Lithophylacii* (Gunther, 1945: 9–10, 320f.). Ray continually bemoaned university and social neglect of natural history (Raven, 1950: 83). Lhwyd had to insist that, though much neglected by society, natural history was a profitable form of learning (Gunther, 1945: 77). Naturalists frequently felt obliged to answer charges that to study minerals was 'base' and trivial. 'I am suspicious', wrote Ray, 'that the vulgar and inconsiderate Reader will be ready to demand, *What needs all this ado? To what purpose so many Words about so trivial a Subject?*' (1732: 167–8).

The devotee of Earth science (as of most sciences) was seen as a rather marginal figure: unusual – but not specially threatening. He conformed to, and reinforced, the image of the man of science as eccentric, maybe a virtuoso (Houghton, 1942); but neither a dangerous, Faust-like magician, nor the archetype of the truth searcher. Though late seventeenth-century pedagogic theories were promoting science, actual education provision made no concessions to study of the Earth. It had no part in a regular liberal career or profession. To investigate the Earth was a particular – and peculiar – decision, which had to be positively chosen, rather than a natural channel into which social conditioning would lead many. Those who chose to pursue it needed to create public interest in their avocation. Of course, social, cultural and intellectual tides were beginning to turn in their favour. Natural knowledge as a projection of reality and a form of culture was rising in public estimation, harmonizing with the values of an age of reason, enlightenment, politeness, toleration, progress and optimism. In due course, society, while not promoting the science, would at length take interest in its findings. Without such social interest the science could not have flourished.

In this section I have examined, rather schematically, some of the corporate characteristics of those investigating the Earth. In most cases we do not know exactly how or why *individuals'* interests were first stimulated. Education had given them potential preparation in the form of medical training, theological knowledge, or general natural philosophical interest. For a few, that preparation led on to more intense inquiry. Thus Burnet's theory emerged out of the philosophical inquiries of the Cambridge Platonists and the age's millennialist expectations, perhaps being sparked off by his crossing the Alps (Jacob and Lockwood, 1972). But most Cambridge Platonists and most millennialists did not write theories of the Earth. Obviously, it is easier to explain interest once controversy had become rife. Whiston's undergraduate days, for example, were spent in the thick of the Burnet controversy, on which he composed student dissertations. Small wonder that he produced his own *New theory* (Farrell, 1973: 22).

## The bridge to the Continent

Before the mid-seventeenth century, there were no distinctively British networks studying the Earth, style of the science, or ideas about the Earth. In other words, British investigation was peripheral to the traditions of Continental Europe. Indeed, throughout the century British men of science travelled widely on the Continent for education and pleasure.[22] As a university elite trained in Latin, they had no trouble in reading Continental books or in personal communication.[23] The establishment of the Royal Society, Continental societies publishing transactions, and the founding of literary reviews towards the end of the century, formalized and extended exchange of information.

Two Continental writers made a particular impact. Athanasius Kircher's *Mundus subterraneus* was influential in the late 1660s as a popular encyclopaedia. Nicolaus Steno was a more challenging stimulus, and perhaps more influential, his *Prodromus* (1669) being 'Englished' by Henry Oldenburg (1671). The *Prodromus* not only sparked off major British controversy on the origin of fossils, but also focused the problem of the relations between fossils, strata and landforms into one of Earth *history* (V. A. Eyles, 1958).

Towards the end of the century, however, the direction of influence was becoming reversed. New Continental works were ceasing to dominate British investigation, while British theories were gaining purchase on the Continent early in the eighteenth century (V. A. Eyles, 1969). Fewer British investigators were travelling abroad – Woodward explained it was because of the 'Commotions which had then so unhappily invaded *Europe*' (1695: 5). The British community was becoming intellectually more self-sufficient. Latin lost favour. Burnet wrote his *Archaeologia philosophicae* (1692) in Latin mainly to prevent the vulgar at home from reading it. There was a gradual orientation away from the international *res publica litterarum* towards a home audience. For nearly a century after 1660, the prevalent mode of fieldwork was fine-toothed observation at home rather than of the grand features of Europe – the Alps, the volcanoes of Italy and the central European mining areas.

# 2

# The natural history of the Earth

## Creating a natural history

The latter part of the seventeenth century was a crucial time in the development of Earth science in Britain. The earlier years of the century had seen traditional fields of discourse increasingly galvanized by the inspiration of Bacon and the activities of the Interregnum. But only later came critical growth of informal and formal social organization, with their new imperatives for scientific practice. Growing numbers, then, were investigating the Earth. This chapter will examine what forms of knowledge they built up. What were the relations between established beliefs, new investigation, and the reconstitution of ideas?

As already discussed, a body of interpretations of the Earth was current, from discrete facts to universal philosophies of Nature. Naturalists themselves proclaimed they were working within frameworks of expectations established by their religious, social, philosophical and metaphysical commitments. Educated Englishmen – and *a fortiori* the group here discussed – were secure in the fundamentals of their philosophy of Nature. The Earth had been created by God as a habitat for man; it had undergone certain stages of development, and would duly be transformed or destroyed by Him. Theirs was an anthropocentric, moral and teleological view of the Earth. They saw in Nature designed order and hierarchy. God controlled the Earth by fixed laws, but also by His immediate Providence. Man should understand the nature and history of the Earth through appropriate methods of inquiry, of which empirical observation was one, but only one. Such inquiry was to be pursued within religious and philosophical frameworks which would also discover natural truth.

Because the Earth was thus so charged with religious, philosophical and human significance, one might assume that the
developing debate about it was shaped solely by such already well-
formulated commitments. But this is only very partially true, and
to overstress this component would be to miss the significance of
the new element in this situation, the re-creation of the Earth as an
object of scientific study. For, at this precise time, consensus as to
the working content (the 'pay-offs') of philosophies of Nature or
natural theology for understanding the Earth survived only at the
most trivial level. Hermetic, Aristotelian, neo-Platonic, Cartesian,
atomistic and organic outlooks upon Nature – and, above all,
eclectic syntheses – vied anarchically for supremacy, but none
achieved hegemony until the triumph of Newtonianism as popularized by his followers and the Boyle Lecturers (Hazard, 1964;
Jacob, 1976).[1]

This warfare among explanatory systems had great impact
upon approaches to the Earth. Where problems of fossils, landforms or minerals could previously be solved within assured philosophies of Nature, the latter part of the century saw the situation
upset. The community possessed no watertight paradigms of
natural explanation, and individuals themselves were often slither-
ing between systems.[2] Evidence from the Earth was pressed into
service in the struggle between warring philosophies of Nature.
It is historically significant that few ventured comprehensive
theories of the Earth before the 1690s, when ascendant Newtonianism could begin to offer a programme of explanation by
simple universal laws.

Between mid-century and the flood of theories in the 1690s,
however, extensive empirical investigation was undertaken, in the
absence of – and to some extent in *response* to the absence of –
generally shared theoretical frameworks. The Earth became conceptualized as a neutral object, distinct from man, to be studied
as God's book awaiting man, its reader. For at this time empirical
investigation, putatively independent of theoretical questions,
became a norm of scientific study, and the ideals of empirical
natural history grew in attractiveness for the study of the Earth.

The writings of the network of naturalists analysed above –
such as Lister, Ray, Plot and Lhwyd – show remarkable unanimity on these matters. Their great desideratum was to perfect

natural history. Listing the chief defects in contemporary natural philosophy, Henry Oldenburg argued:

And, I think, I may say, that a *Natural History of Countries* is most wanting; which, if well drawn, would afford us a copious view, and a delightful prospect of the great variety of Soyls, Fountains, Rivers, Lakes, etc, in the several places of this globe...(Oldenburg, 1675–6: 552)

Understanding of Nature, they agreed, was still being hindered by over-bold speculation and hypotheses, and by credulity. Lhwyd hoped Woodward's *Essay* (1695) would attract such ridicule that it would 'make men preferre Natural History to these romantic theories' (Gunther, 1945: 269). Yet he needed to repeat himself on the appearance of Whiston's *New theory*: 'I hope that ere long we shall have this wrangling philosophy laugh'd out of countenance, and ye plain Natural History better esteem'd of' (Gunther, 1945: 283).

Ray was notably severe against credulity, the antidote to which was observation:

As for what Dr Plot produces out of Camden and Childrey, in confirmation of his fourth argument, viz. that the *Ophiomorphites* of Cainesham have some of them heads, I doubt not but it is a mistake, proceeding from their credulity. For Mr Willughby and myself inquiring diligently there after such stones, the common people affirmed that there were such found; we not satisfied with their assertion, but desirous ourselves to see them, were at last directed to a man's house who was said to have one, to whom when we came, he showed us the stone, which indeed at the upper extreme had some kind of knob or protuberance of stone, but not at all resembling the head of any animal. Such a kind of stone might, perhaps, be shown to Mr Camden, whose fancy being possessed with the vulgar conceit, he might without any strict view or examination of it, admit it to be what the vulgar would have it. (Lankester, 1848: 155)

They praised history as noble as well as useful. As Hans Sloane saw it,

Knowledge of Natural-History, being observation of Matter of Fact, is more certain than Others, and in my slender Opinion, less subject to Mistakes than Reasonings, Hypotheses and Deductions are. (de Beer, 1953: 96; cf. Gunther, 1945: 78, and Woodward, 1695: 5–8)

And most important, they tried to lay down canons for its proper practice. They demanded exact, first-hand data as the keystone of the science. Furthermore, they became personally aware, from experience, of the extreme practical difficulties posed by the complexity of Nature. Lhwyd wrote

Ye frequent observations I have made on such bodies [fossils] have hitherto affoarded litle better satisfaction, than, repeated occasions of wonder and amazement, for as much as I have often (I may say continually) experienced, that what one days observation suggested, was by those of ye next calld in question, if not totally contradicted & overthrown. (Gunther, 1945: 381)

Experience taught him that thoroughness was a prerequisite for fossil studies. Explaining why he needed seven years to complete a projected natural history, Lhwyd insisted, 'with Natural History know there's no good to be done in't without repeated observations' (Gunther, 1945: 270). Before embarking upon his natural history of Wales, he reflected

It's well known, no kind of writing requires more expenses and Fatigue, than that of Natural History and Antiquities; it being impossible to perform any thing accurately in those Studies, without much travailling, and diligent Searching, as well the most desert Rocks and Mountains, as the more frequented Valleys and Plains. (Gunther, 1945: 14)

As he later explained about fossils, 'I find that variety of specimens from several places are of more use in this part of Natural History than any other' (Gunther, 1945: 275).

The natural history project had powerful functions. It projected a conception of the Earth as capable of successful investigation, because an ensemble of discrete, stable, real objects. As Sir Hans Sloane wrote,

There are things we are sure of, so far as our Senses are not fallible: and which in probability have been ever since the Creation and will remain to the End of the World in the same condition we now find them. (de Beer, 1953: 96)

Thus their commitment to Baconian natural history was not vacuous rhetoric. It deeply influenced their practice. Their letters show them being baffled by new phenomena; changing their minds, unable to reach just conclusions. Furthermore, they were pioneering new techniques to cope with the continuously emerging problems of a natural history of the Earth.

As already noted, limitless travel was one of Lhwyd's solutions. Others sought to develop artificial techniques. The aim for a new representation of the Earth's crust crystallized in a demand for physical maps. Thomas Burnet complained that maps showed 'things of chief use to civil affairs and commerce. . .not Philosophy or Natural History' (Burnet, 1684: 140). John Evelyn brought to the Royal Society's notice some three-dimensional maps produced in France 'in low relievo', as attempts to embody relief in artificial form (1665: 99–100). Martin Lister and John Aubrey further sought mineral maps of Britain to tabulate outcrops, soils and mineral resources. 'I have often time wished', wrote Aubrey, 'for a map of England coloured according to the colours of the Earth, with markers of the fossils and minerals' (Gunther, 1925: 214; M. Hunter, 1975: 113). Lister spelt out his vision in greater detail, in the *Philosophical transactions*. Despite its loose ends, Lister's plan had clearly grasped the diversity of outcroppings of distinctive rocks, and that a map representing these could *predict* structure and mineral deposits elsewhere (1684b). That no such maps were actually drawn up for almost a century was a result partly of continuing ignorance of the precise details of exposures across much of Britain, and partly also of natural history's strong predisposition to conceive of the Earth's surface in a pointillist manner as an accumulation of individual deposits rather than in terms of masses and structure (North, 1928).

Other visual aids could make more headway. Some were verbal and tabular, such as vertical columns of the order and depths of strata as deduced from wells and mine shafts (e.g. Bellers, 1712). Some were pictorial, such as bird's-eye sketches showing land-slips, or horizontal panoramic cross-sections of the strata, as pioneered by John Strachey (1719; 1725). The use and sophistication of pictorial representations developed. The series of articles in the *Philosophical transactions* on the Giant's Causeway moved from the purely verbal (Bulkeley, 1693), to one with a primitive sketch (Foley, 1694), to a more sophisticated and accurate representation (T. Molyneux, 1697). Fossil identifications above all were newly aided by high quality engravings (Hooke, 1665; Lister, 1674).

Lastly the questionnaire, circulated by naturalists such as Plot (1677), Lhwyd (1909) and Woodward (1696), became another characteristic weapon in the armoury of the natural history project. Despite difficulties in obtaining replies (Gunther, 1945: 347), they were circulated with remarkable assiduity (Lhwyd distributing 4000 for his survey of Wales).[3] Their format was developed out of the 'Articles of Inquiries touching Mines' and the 'General Heads for a Natural History of a Countrey', published in early numbers of *Philosophical transactions* (1666c; 1666a), and like them concentrated on accumulating discrete data about physical geography and mineral objects:

Among these *Subterraneal* observations may be taken notice of, what sorts of Minerals of any kind they want, as well as what they have; Then, what Quarries the Country affords, and the particular conditions both of the Quarries and the Stones; As also, how the Beds of Stone lye, in reference to North and South etc; What Clays and Earths it affords. (*Phil. trans.*, 1666a: 189)

Questionnaires also grew in sophistication. Woodward, for example, requested not merely information on fossils and minerals, but also details of their location, material and position in the strata. He specified equipment for making observations, and gave precise instructions for preserving, labelling and packing of specimens (1728: 93f.). But the sophistication lay wholly within the terms of the project of natural history.

The techniques and intellectual frameworks of these naturalists were, then, geared to natural history fieldwork. Lhwyd's seminal *Lithophylacii Britannici ichnographia* (1699) was designed both physically and intellectually to be carried on field expeditions (Gunther, 1945: 153). They were not doing 'geological' fieldwork, as later generations were to understand it, but were rather compiling a natural history of regions of the Earth, or the natural history of the mineral kingdom. What were the intellectual assumptions embodied in this enterprise?

One was a long-established tradition of animal, vegetable and mineral classification which related individual specimens to a putatively universal, timeless and continuous scale of Creation. The main tasks within this tradition were to identify particular specimens, to accumulate a total roll-call of Creation, to discover essential characteristics and differences for classifying into natural, or artificial, groupings; and to provide exact verbal and visual descriptions to aid future identification.

Lister, Ray, Lhwyd and many other experts on minerals and extraneous fossils subscribed to this ambition of understanding terrestrial objects by locating them within a universal explanatory reference grid. Questions of the historical, local and material relationships of rocks and fossils to their physical matrices were secondary. Thus Lhwyd characterized his *Lithophylacii* (1699) as not 'a satisfactory account of the origin of fossil shells', but rather 'a classical enumeration of what Form'd Stones my collection consists of' (Gunther, 1945: 299). Yet while the seventeenth-century natural history of the mineral kingdom sought different explanations from later geology, it marked an important stage in investigation. Without the descriptions, identifications and classifications of Plot, Lister, Ray, Webster, Lhwyd and Woodward, future generations would have built on sand.

The other focus of natural history was regional – at this stage, almost always one county, or county by county, though the aim was to cover the whole country, Plot for example calling his *Oxfordshire* 'an Essay towards the *Natural History* of England'. This was an outgrowth of earlier antiquarian and topographical traditions, perhaps influenced by Boyle's 'General heads for a Natural History of a Countrey' (*Phil. trans.*, 1666a). That a county (i.e. a human not a physical area) was still chosen as the object suggests this aetiology, as also the calculated gentry market. Some were pure compilation, as was Childrey who in his *Britannia Baconica* (1661) merely listed, higgledy-piggledy, all he knew about each county, *seriatim*. Some employed the timeless natural historical grid just described, showing little interest in physical relationships. Plot thus adopted Bacon's scheme, as modified by Boyle, by dividing his work into three main parts: 'Natural Things, such as either she [Nature] hath retained the same from the beginning, or freely produces in her ordinary course. . .Secondly, her *extravagancies* and *defects*. . .And then lastly, as she is restrained, forced, fashioned, or determined, by Artificial Operations' (Plot, 1677: i). He then subdivided it into discourses on animals, vegetables and minerals, in turn further broken down into parts according as each related to the elements air, water and earth (Plot, 1677: 81). Thus, for example, fossils were treated not *en masse* in relation to their strata matrix, as was to become common, but distributed throughout the work according to whether they were believed either the regularities or irregularities of Nature, and by whether they resembled the products of air, water or earth.

Once again, as with classificatory natural history, it would be easy but superficial to overlook the real historical connections between early pioneering county natural histories and later geology. Yet approach and technique were under continual revaluation, as may be evident from Aubrey's searching for the right *name* for the genre – and ending with 'Chorographia super et subterranea naturalis' (1847: Preface; M. Hunter, 1975: 113f.). Before 1660 county histories were not predominantly concerned with natural history. Plot's works were then divided between natural and civil history, but Thomas Robinson's *Westmorland and Cumberland* (1709) and Morton's *Northampton-shire* (1712) were overwhelmingly natural historical in their orientation. Similarly, Camden's *Britannia* (1586) had relied very heavily upon scholarly sources for its natural history. Childrey's significantly titled *Britannia Baconica* (1661) made considerable use of first-hand observations for the limited parts of Britain with which he was familiar. With Plot, the fine-toothed county natural history appeared, including maps and plates, built upon the findings of questionnaires and personal observation, conforming to an organized programme. Plot was less impressed with marvels and the bizarre than Childrey (Paffard, 1970).

By the time of Morton's *Northampton-shire* (1712), the framework for interpreting a region had altered considerably. Morton had out-Plotted Plot and visited all the villages in his county. He had strict, explicit, methods of inquiry.

Most of the Observations that this Second Chapter consists of have been made upon the Interiour Parts of the Earth expos'd to view in the *Quarries*. My Method in taking them had been usually this. First I noted the *Species*, *Thickness*, *Order* and other Circumstances of all the *Strata*, beginning with the uppermost and so proceeding downwards to the Bottom of the Pit; particularly the several *Kinds* and Varieties of *Stone*, the *Uses* of Each, and how each is *affected* or wrought upon by *Water*, *Frost* or *Heat*; the *Number* of the *Strata* into which by means of *Horizontal* and *Parallel Fissures* it is divided; and the *Eavenness* or *Inequality* of the Surface of the *Stratum*; As also the *Foreign* Matter and *Heterogeneous* Bodies, especially the *Sea shells*, enclosed in the *Stone*. (Morton, 1712: 97)

He drew on earlier British local natural history works and Lhwyd's *Lithophylacii* for reference and comparison. He applied, tested and confirmed some of Woodward's ideas about stratification (1712: 98). More importantly, his interests and perspectives differed considerably from Plot's – a shift which in part registers new horizons opened by his fieldwork. Having organized his material after the universal classificatory scheme suggested by Bacon and Boyle, Plot had not seen fit to offer a general introduction to the physical geography of Oxfordshire or Staffordshire.

Morton, however, programmatically introduced his work with an important discussion of the position of Northamptonshire within the physical structure of Britain. He then presented integrated accounts of the county's relief, relating this to strata and topography. In Morton's work, local natural history was initiating the more structural, proto-geological topography which developed in the eighteenth century.

A parallel instance of the developing inner logic of natural history would be the series of articles appearing in the *Philosophical transactions* upon the Giant's Causeway – revealing new techniques, problems and perspectives. The first, by Sir Richard Bulkeley (1693), merely reported meeting a man who had visited the Causeway. He requested a list of Queries from Martin Lister for when he should personally observe it. Understandably, Bulkeley's description was imprecise. Samuel Foley (1694) suggested answers to certain issues raised by Bulkeley, and included a map and a drawing of the Causeway made by a local collector. In his comments on Foley's paper, Thomas Molyneux (1694) condemned Bulkeley for his slapdash account, and programmatically attacked the 'vulgar' for believing that the Causeway was an artefact of giants or fairies. He confessed, however, that he had not yet seen it, but nevertheless pointed out that its columns resembled nothing so much as the '*Lapis Basaltes Misenus* described by *Kentmannus* in *Gesner de Figuris Lapidum*, from whence *Boetius* takes both his Figure and Description' (181). He also speculated that hexagonal columns might by the analogy of nature be huge forms of astroites and entrochi.

William Molyneux's contribution (1697a) was a well-keyed sketch of the Causeway. Thomas Molyneux's second account (1698) praised his brother's drawing as far superior to Foley's earlier one (as is indeed the case), and elucidated it. He still had not had time to inspect the Causeway himself, but had assiduously collected information about it from acquaintances. Though its origin was an important problem, it best befitted men such as himself and the present state of philosophy to describe it rather than to concoct hypotheses. By 1699 Lhwyd could write to Tancred Robinson that the Causeway was 'so well described in the *Phil. Transact.* that nothing can be added to that account of it', and used these accounts to point the comparison that 'we have the same stone on the top of Cader Idris' (Gunther, 1945: 420).

The continuing limitations of the late seventeenth-century tradition are plain. Where thorough, it was local and narrow. Where ambitious, it was shallow. It had developed greater capacity to collect data than to use them. Nevertheless a programme had been launched which earnestly laid the Earth open for inventory and analysis, and naturalists had internalized strong imperatives to anatomize the globe with humble rigour. This proved a favourable combination for major reconceptualizing of the Earth, both as a whole and seen as distinct problem areas, as I shall now examine. In the succeeding analysis I have to some extent followed convenience, but much more so respected perceived contemporary boundaries of knowledge. For though distinct sciences of geomorphology, palaeontology and stratigraphy were not yet conceived, contemporaries themselves organized in different compartments their knowledge of the static and the dynamic, the whole and the part, life and rocks, past and present. Geology is when these come to be integrated, and that is not till the end of the eighteenth century.

### Landforms and habitat

The most immediate aspects of the Earth are its face and surface processes. Up to the mid-seventeenth century – and frequently beyond – present landforms were most commonly interpreted as remnants of a perfect, original Earth, sculpted at the Creation; a view made more plausible by almost universal acceptance of a short time-scale. The 'essential' features of relief – the division of the globe into land and sea, hills and rivers, valleys and lakes – were thought coeval with the Creation. Since then, the habitat had been modified, but not unrecognizably transformed, by a combination of one or two great – perhaps miraculous – catastrophes and by regular change, which were collectively interpreted as 'decay'. 'Decay' included such processes as coastal erosion, land denudation, extension of deserts and other uninhabitable places, eruption of volcanoes and earthquakes. 'Decay' brought malfunctioning and ugliness. It was both the historical condition of the Earth and a continuing curse (Davies, 1966a; 1969: chs i, ii).

Such a way of assigning to aspects of landforms their distinct origins and epochs embodied the Christian apocalyptic synopsis of human and cosmic history as corruption induced by the Fall. It also reinforced Classical belief in natural cycles, the organic analogy, and nostalgia for a pristine golden age (Tuveson, 1949: ch. ii; Jacob and Lockwood, 1972). It fed fascination with freaks and marvels in Nature, and the allure of the grotesque. Earthquakes and volcanoes, bottomless pits, intermittent springs, erratic boulders, ships and shells found inland, tidal waves, rivers that changed their course – all were staples of popular literature, evidence that Nature was crazy, sick, in decline. It was a perception characteristic of a system of economy and society in which man was essentially tied to the soil, and the fate of man seen to be inextricable from that of his planet, his habitat. The economy of the Earth was not conceived as an autonomous self-existent system, nor man as an independent agent.

This natural philosophy of decay stimulated interest in aspects of the Earth's appearance and processes such as earthquakes and volcanoes, denudation, shifting coastlines, complex water systems (Biswas, 1972; Tuan, 1968). It encouraged close study of change, and of the analogies between present changes and past. It buttressed speculation about growth of minerals and metals. Davies has analysed emphasis on changing landforms in Renaissance thinkers (1969: chs. i, ii), but *Philosophical transactions* papers right through to the end of the century show the continuing vitality of the tradition. Many pinpointed shells and anchors found inland, or subterranean forests, as evidence of shifting coastlines. Gigantic bones found underground likewise attested major catastrophes of the land. Landslips, pot-holes, sinks and undermined cliffs were investigated.[4] Controversy raged whether an isthmus had once joined Britain and France, swept away either by a sudden débâcle or by gradual attrition (*Chartham News*, 1701; Wallis, 1701). And a score of papers recorded and debated earthquakes and volcanic eruptions (e.g. Pigot, 1683; Lister, 1684a). Especially in regard to theoretic understanding, Davies is right to locate a growing 'denudation dilemma' (1966a; 1966b). But within local empirical research, the problems at hand continued to indicate an Earth undergoing change (e.g. de la Pryme, 1869: 67). Nevertheless, the wholesale past and present changes formerly so eagerly accepted were gradually coming under question. Instrumental in this shift was the programme of natural history observations, but so also were profound conceptual realignments.

The impact of the new framework of natural history should not be minimized. Staking its credentials on rigour, it combatted credulity and unsubstantiated claims, and demanded first-hand observation and measurement. It was pleased to confront traditional accounts of intermittent rivers and islands, bottomless pits, shifting coast-lines, mountains like Mam Tor that poured forth endless materials without ever diminishing in height – tales which had portrayed the Earth in decay – and to burst them like bubbles. Penpark Hole was by tradition bottomless and haunted. Sir Robert Southwell, founder member of the Royal Society, measured it, and found it thirty yards deep, 'which does much lessen the credit and terrour of this hole' (1683; cf. Davies, 1969: 42). Thus to demystify the globe gave natural history a victory, and also extended the territory of disenchanted Nature upon which natural history could operate.

The same scepticism, used to annexe the right to reinterpret intellectual territory, confronted contemporary processes. Within human testimony, the outward face of Nature had remained extraordinarily constant. The map of Europe had scarcely altered since the Greeks. This meant, as Lhwyd recognized, that belief that the Earth had substantially altered its physiognomy was only plausible if one accepted an extremely long time-scale – against which he set his face (Lankester, 1848: 241). For close inspection suggested that only marsh and sea-shore had been formed since the Deluge, not consolidated rock; mountains were not being visibly denuded; land and sea were not changing places. There had been merely marginal landscape changes since the Deluge, but no real trend or revolution. As John Woodward asserted, his observations in England, 'the far greatest part whereof I travelled over on purpose to make them' (1695: 3), led him to conclude:

although there do indeed happen some Alterations in the Globe, yet they are very slight and almost imperceptible, and such as tend rather to the benefit and conservation of the Earth and its Productions, than to the disorder and destruction both of the one and the other, ...the terraqueous Globe is to this day nearly in the same condition that the Universal Deluge

left it; being also like to continue so till the time of its final ruin and dissolution, preserved to the same end for which 'twas first formed and by the
same Power which hath secured it hitherto. (1695: 46–7)

But indeed, as Woodward's own thought makes clear, deep
conceptual currents – religious, philosophical and aesthetic – of
which the imperatives of natural history were in some degree an
expression, were also displacing the theory of the Earth in decay.
In Restoration England fundamentalist Calvinist Protestantism
was being hounded out by a theology which strove to establish
order and stability – moral, social and natural – through stressing
God's love, wisdom and rationality. For the Earth, this entailed
re-affirmation of design and control. Lawful functioning attested
the wisdom of God's original order and His pre-ordained harmony of structures and functions. Not marvels, miracles or catastrophes, but the perfection of His creation best sang God's praises
(Ray, 1691; Bentley, 1692–3).

These pressures compelled major reconceptualization. Mountains, till recently considered excrescences, were now re-envisaged
as objectively functional in their own right, essential to the
efficiency of the Earth-machine. 'But here it may be objected',
Ray confessed,

That the present Earth looks like a heap of Rubbish and Ruines; and that
there are no greater examples of confusion in Nature than Mountains
singly or jointly considered; and that there appear not the least footsteps
of any Art or Counsel either in the Figure and Shape, or Order and Dis-
position of Mountains and Rocks. Wherefore it is not likely they came so
out of Gods hands...To which I answer, That the present face of the Earth
with all its Mountains and Hills, its Promontaries and Rocks, as rude and
deformed as they appear, seems to me a very beautiful and pleasant object.
(Ray, 1692: 165)

This natural theology of stable order entailed the essential
powerlessness of natural terrestrial forces. Thus John Woodward's
*Essay* (1695) repudiated ancient geographers' conclusions that the
Earth's surface was a great theatre of change, allowing 'those
ancient pagan writers' were 'much excusable as to this matter'
because they 'had only dark and faint Ideas, narrow and scanty
Conceptions of Providence' (p. 56). Especially culpable, thought
Woodward, was the Epicureans' theodicy, for, believing that 'the
world was framed by Chance', and that 'either there was no God
at all, or, which is much the same thing, that he was an impotent

and lazy *Being* and wholly without concern for the Affairs of this lower World', they concluded that 'the Elements were at constant Strife and War with each other: that in some places the Sea invaded the Land: in others, the Land got ground of the Sea, that all Nature was in a Hurry and Tumult; and that as the World was first made, so should it be again dissolved and destroyed, by Chance' (p. 58).

In short, because to see the Earth in continuous change or decay undermined God, it dissolved intellectual order, and so the possibility of science. Woodward's outlook invoked the mechanical philosophy's denial that Nature was self-existent. Nature was the will of God, and matter was passive and inert unless informed by God's will (cf. Hooykaas, 1972: ch. v).

Aesthetic expectations were transformed in a parallel way as the new cosmology affirmed the independent objectivity of Nature. Wild scenes of Nature – mountains, gorges, cliffs, great oceans – earlier considered disgusting, were now gaining legitimacy as aesthetic experience; not beautiful (cf. Halley, in Armitage, 1966: 136) but sublime, and certainly not warts, blemishes and disfigurements upon Creation. Earlier indifference to grand scenery was beginning to yield to active involvement in the aesthetic experience of Nature, as human subserviency to environment gave way to dominance. The 'aesthetics of the infinite' were being born (Nicolson, 1959: chs. i, ii).

So the programme of observations, and a new cosmology, united in a view of the Earth as a stable, well-designed independent, objective mechanism, in which 'the World continues still in as good state and condition as it was two thousand years ago, without the least impairment or decay...and therefore arguing from the past to the future, it will in all likelyhood so continue two thousand years more' (Ray, 1692: 175). This 'restoration' of the Earth was in the air throughout the Restoration age, but with the triumph of the Newtonian world-view in Woodward's *Essay* (1695) and Whiston's *New theory* (1696) it found its classic embodiment. In closing doors on some aspects of the interpretation of the Earth, it opened others. Legitimation of the functions of phenomena previously thought imperfect made possible systematic study of mountains, and stimulated inquiry into terrestrial forces understood as *systems* – pre-eminently the hydrological cycle – with detailed attention for the first time to the problem of the origin of rivers (Biswas, 1972: ch. x).

## Meanings of fossils

Growing investigation of fossils similarly involved the carving out of a practice of close and rigorous examination, one no less resonant with deep issues of how Nature and the Earth were constituted. Fossils, however, more sharply exemplify two crucial developments. Firstly, the forging of basic techniques of observation, recording and comparison of information was a *sine qua non* for opening up a whole sub-discipline of identification, pitching interpretation onto a more subtle plane. Secondly, in debate on the nature and origin – the 'meaning' (Rudwick, 1972) – of fossils, all intellectual consensus had broken down – if it had ever existed. To lead them out of the chaos of conflicting interpretations of the nature, origin and significance of fossils, naturalists increasingly built up the expectation that turning to Nature, to ever more minute and comprehensive examination of fossil objects, *in situ* and under the microscope, would settle the issues. In the event – and more blatantly than with landforms – the natural history project did not provide 'solutions'. Resolution emerged from deeper thought constellations which were forming. It did however initiate programmes of investigation which had profound momentum beyond their original ends.

Throughout this period, the word 'fossil' meant anything 'dug up', any substance found in the ground of some special interest (cf. Carr, 1974). Furthermore, up to mid-century, there was no distinct British study of 'figured stones' (i.e. those 'fossils' with appearances similar to creatures and plants). This was partly because strict classification of any mineral substances had hardly yet become important, but also because such figured stones were not regarded as a privileged domain of Nature. After all, the Great Chain of Being emphasized the *continuity* of Nature's three kingdoms. Stones, minerals and crystals all grew into distinctive forms. Only a minority of naturalists believed that most or all figured stones were genuinely 'extraneous' – i.e. the reliquiae of organisms. Indeed, the problem of petrifaction attracted more attention than did figured stones as such, and though petrifaction included figured stones, its main loci were the nature and origin of mineral objects in general, in macrocosm and microcosm (e.g. gall-stones), stalactites, sparry concretions, crystals and the root problems of solidity and cohesion (*Phil. trans.*, 1666b; Boyle, 1672; Sherley, 1672). In Restoration England, however, figured stones steadily became isolated as an area of debate. Not that all naturalists now began to recognize them as organic remains; on the contrary, their origin remained a problem – though still not the paramount one. The main development was the newly intensive observation, description and analysis of figured stones.

The value to succeeding generations of the groundwork of Hooke in microscopic investigation of fossil objects, and of Lister, Ray, Beaumont, Plot and Lhwyd in collecting, illustrating and collating, cannot be exaggerated. Their books, correspondence and *Philosophical transactions* papers show the scope of their work (cf. Ray, in Lankester, 1848: 212f.). Specimens needed to be identified against earlier authorities, often themselves imprecise and without illustrations. Identification was boosted by the development of exact fossil illustration, as in Hooke's *Micrographia* (1665) and *Posthumous works* (1705), Lhwyd's *Lithophylacii* (1699), and Lister's *Philosophical transactions* articles and his *Historia conchylorum* (1686–92). But many specimens were nondescripts – for example, the fossil plants which particularly puzzled Beaumont (1676; 1683). Others seemed to exist only in fragments. Amongst these, glossopetrae, which exactly matched modern sharks' teeth, did not create great difficulties (Ray, 1738: 267). But trochites, entrochi and astroites were problems of a different order, since there was no agreement whether they more exactly resembled the spines of sea urchins or plant stalks, and in neither case was the correspondence with present species exact (e.g. W. Nicolson, in Nichols, 1809: 83f.). Of astroites Lister confessed, 'I pretend not to discover to you their Original no more than I did of the *Entrochi*' (Lister, 1674–5b: 274). Quite irrespective of their *origin*, the issue of distinguishing between those figured stones which could be classed as mere mineral forms, such as crystals, those which *resembled* plants, and those which *resembled* animals now became pressing, if intractable, in the work of Ray, Lister and Lhwyd (Gunther, 1928: 215–16).

Yet strides were made. For the naturalists studying fossils were precisely those most involved in identifying living specimens. Lister was perhaps the world's authority on shells, Plot on snails and spiders, Ray, 'the Linnaeus of the time', had a synoptic classifying vision (Dance, 1966; Raven, 1950: 452). By the beginning of the eighteenth century, in the work of Woodward and especially in Lhwyd's *Lithophylacii*, all known figured stones could be assimilated into a working descriptive classification.

Two other advances in the mechanics of investigation were important. This period saw the foundation of two distinguished public fossil collections (the Royal Society's and the Ashmolean's), and many important private ones (above all those of Cole, Lister, Beaumont and Woodward). Two of these – the Ashmolean and Woodward's – served as the basis for published fossil catalogues subsequently much used (Gunther, 1945: 89). Collections were essential for future identification and exchange of specimens, and for the comparison of the fossil with the living world.

Secondly, the natural history method indirectly encouraged punctilious recording of the lithology of figured stones and their position in the general order of strata and relief.[5] The practice of noting the rock basis was not initially intended to solve the problems of origins, but rather to fulfil the Baconian injunction of maximizing information about natural objects. However, major naturalists did use this information to argue – diverse – cases for origins.[6] More importantly, however, within the growing fabric of Earth science, this information was used in later years to support interpretations of fossils and strata of which its recorders had no conception. Thus, in the 1730s, on the basis of just such data, John Rawthmell advanced the pregnant suggestion that fossils were pre-eminently found in England on an axis from the Cotswolds through to the North Sea (R. Richardson, 1835: 316). 'Plot's Oxfordshire, Morton's Northamptonshire, and Woodward's Catalogue of fossils became his favourite study', wrote John Phillips about William Smith, 'and probably no other reader has drawn from those curious but neglected works so many valuable geological data' (1844: 24).

Certainly, recording information on individual figured stones and their rock-matrix began to provoke larger questions of their physical and temporal relationships – as in Hooke's comparison between the antiquary discovering buried coins and the naturalist finding fossils (Hooke, 1705: 335; 411). Awareness grew of the puzzling fact that some rock-types and localities were highly fossiliferous while others were barren, and old misapprehensions were broken down, such as Ray's early belief that fossils were not found on high mountain sides.

It is not that there was indifference to the origin of figured stones. Rather, many naturalists who were trying to prove a case by the use of exact empirical data became aware that the data available to them did not resolve the issues. They found their newly-discovered facts, on which they had pinned their faith, did not conclusively tip the balance in favour of one theory. Indeed they became aware – almost despite themselves – that the *meaning* of particular facts appeared different within different theories; that facts required interpretation and then necessarily from within theoretical frameworks. And so they resolved their dilemmas largely on architectonic, metaphysical convictions. While these beliefs were of course crucial and profound to them, such decisions undercut their aim of constructing certainty by empirical research, and – because of the incommensurability of metaphysical systems – failed to convince others, or themselves. Plot, Lhwyd and Ray all showed startling vacillations (Butler, 1968: 150–9).

The issue was not whether figured stones as such were essentially organic or mineral, since certain figured stones were evidently mere minerals (e.g. rocks which 'accidentally' had the outward form of, say, a human head) and others obviously organic remains (e.g. buried vertebrate bones found in an unmineralized condition). The issue was rather *which* fossils had *which* origin – it was one of selective classification. The most popular theory was that most figured stones were *lapides sui generis*, natural mineral growths formed by some kind of virtue, petrifying fluid, mineral juice, or stellar influences. It had positive and negative support. Positively, it was plausible because such growths were integral to the Classical–Renaissance Aristotelian, Platonic and Hermetic philosophies of Nature which held sway in mid-century, and to routine mineral chemistry. Given that stones of many kinds manifestly grew *simpliciter* (minerals, metals, stalactites, gall-stones), and certain stones grew into defined geometrical forms (e.g. crystals), why should not others, analogously, grow into vegetable and animal forms? The Great Chain of Being seemed to require some intermediate, indigenous entities between vegetables and inert minerals. Figured stones might be analogous to coral, as rock which grew.

Furthermore, empirical investigations confirmed this view at many points. Many figured stones – as close observation or the microscope revealed – differed in form from current species and hence were not indisputably relics of once-living creatures. If they were, this raised the spectre of extinction – which almost no writer found likely in Nature, or religiously and metaphysically accept able (Ray, 1693: 72; Gunther, 1928: 260). Again, inquiry found figured stones generally uniform in material with their matrix rock, which suggested they were growths integral to the bedrock, rather than extraneous deposited bodies. In addition, correlations of occurrence were discovered between particular figured stones and particular matrix rocks, which could imply that the chemistry of certain rocks engendered fossil stones, whereas others were barren. If fossils had been organic remains, they would, it was argued, have been more randomly distributed (e.g. by the Deluge) among all rocks.

Furthermore, powerful objections were paraded against the organic remains theory. The trump card constantly played was that no known mechanism could have infiltrated creatures, or their remains, into solid rocks; nor was there any occasion when this could have happened. The Noachian Deluge was ruled out, for Scripture apparently described a brief and quiet event; and likewise the Biblical Creation, because Genesis stated that dry land had appeared before the creation of life. No other recorded event nor actual process could have buried the remains of crea tures so deeply within rocks, or so high upon mountains.[7] Small scale petrifaction of shells in estuaries and limestone streams was not an adequate analogy. Furthermore, a plague of additional physical and metaphysical problems attended the organic remains theory: the likelihood of extinction; the problem of why fossils were found thousands of miles away from their present habitat; of how petrification occurred; of why certain species' remains had been plentifully preserved, and others' not at all.

Though the interpretation of figured stones as natural objects predominated up to the 1690s, it did not go unchallenged by naturalists such as Hooke and Ray. Their strategic argument that fossils were organic remains hinged on the uniformity of Nature. For unpetrified bones and shells were to be found in unconsolidated sediments, peats, mosses and alluvium. Semi-petrified bones and shells lay in rather more consolidated beds. Human artefacts likewise could be found at various depths in course of transformation from their original condition, coated with, or changed into, the surrounding rock materials. Likewise, buried wood and tree trunks had been found in all conditions, from that of the living specimen through to complete petrifaction or coal. The formation of sparry concretions in limestone streams was similar. Had not bones and shells as well, then, been buried and gradually transformed – by heat, chemical action, or pressure – into figured stones?

Mechanisms for this were proposed. The Noachian Deluge was one – particularly favoured by those who assumed that figured stones mainly lay superficially. Woodward, who recognized the depth of figured stones, needed to develop a theory in which the Deluge had totally dissolved all land, to explain the entombment of fossils. Hooke, for his part, tried to show how organic remains had been consolidated into beds gradually forming on the sea floor, and then raised to form dry land by the action of earthquakes and volcanoes, and perhaps by shifts in the Earth's centre of gravity. Hooke however acknowledged the speculative nature of his theory, and that it shook common belief in the novity of the Earth.

Hence broader beliefs were needed to reinforce interpretation of fossils as organic remains. Many figured stones *were* identical in form to living plants and animals (Hooke, 1705: 281–5). Nature would be acting irrationally, unpurposively, if she had created those structures without life, without functions. If figured stones were *lapides sui generis*, what possible purpose had they in Creation? Such a belief would bring teleology into disrepute and undermine natural philosophy. Above all, why should Nature have produced *fragments* – bodies resembling teeth and individual bones?

A whole generation of naturalists was deeply perplexed on the question of the nature of fossils, and few felt confident that data or philosophy clinched the issue. Some like Lhwyd developed entirely alternative explanations (Huddesford, 1760). Plot weighed up both sides at length:

*Whether the stones we find in the forms* of Shell-fish *be* Lapides sui generis, *naturally produced by some extraordinary* plastic virtue, *latent in the Earth or Quarries where they are found? Or whether they rather owe their form and figuration to the* shells of the Fishes *they represent*, brought to the places *where they are now found by a* Deluge, Earth-quakes, *or some other such means.* (Plot, 1677: 111)

Lhwyd similarly puzzled himself,

whether I conclude with ye general opinion, that they have been reposited in the places we find them at ye universal Deluge and so preserved to our time; or that they are original productions of Nature, there form'd from some plastic power or salts or other minerals. (Gunther, 1945: 391)

'For my part', wrote Ray, 'the difficulties seem insuperable' (Gunther, 1928: 278). Yet the commitments were not random. The chief natural historians were disposed to conceive figured stones as natural mineral objects. For the natural history method focused their attention on the visible form and structure of fossils, and on problems of their identification and classification, rather than on origins and questions of Earth history. Furthermore, as expert naturalists, they were impressed by the real distinctions of form between many figured stones and actual species. Their credentials depending heavily upon rigour, they dismissed as fanciful notions that organic remains had been lodged in rocks. By contrast, those with antiquarian connections, and those approaching the Earth from historical perspectives, tended to see them as organic remains, 'medals of creation', buried by catastrophes or by Time, analogous to human artefacts (Hooke, 1705: 321; Schneer, 1954).

This conflict, however, was achieving some resolution by the 1690s. Greater empirical awareness developed of the large vertebrate remains found in various degrees of petrifaction in many parts of the globe, not least in Ireland and Siberia (T. Molyneux, 1697; Tentzelius, 1697). These were emphatically not mineral objects in their own right. But the revolution came from outside, from radically shifting commitments to philosophies of Nature, and from attempts to forge a total interpretation of Earth history amongst the generation of theorists in the 1690s. By the end of the century the retreat of Classical and Renaissance philosophies of natural history before the mechanical philosophy had become a rout, especially once the latter received the blessing of Newtonian matter theory and celestial physics. With their demise, theories of figured stones geared to metaphysical foundations such as the creative and sportive powers of Nature, the continuity of Nature, the virtues and tendencies of the mineral world, lost their main props. Whereas entombment of species had previously appeared most fanciful, now the supposed creative powers of Nature threatened the straightening norms of scientificity.

Moreover – and the two trends are related – the 1690s saw growing attempts to construct all-embracing theories of the Earth. The Deluge played a deep structural role within all of them. If seen as organic remains, figured stones provided telling confirmatory evidence for this Deluge; while at the same time, reaffirmation of the Deluge buttressed the view that formed stones were buried organic remains. Even though no new convincing explanatory mechanism was offered, Deluge theory and the organic remains explanation of fossils undoubtedly marched hand-in-hand into the eighteenth century. The *lapides sui generis* view had not been 'refuted' – it is no more 'refutable' than phlogiston – nor even come under steadily mounting empirical fire. Rather, its whole explanatory framework had become obsolete (Butler, 1968: ch. iv). By the mid-eighteenth century it could be a joke (Anon., 1947).

Fifty years' investigation constituted fossils as a major object of natural history inquiry, establishing a tradition of meticulous examination and classification. Natural history methods alone, however, bequeathed unresolved interpretative dilemmas. The need to establish a historical theory of the globe in the 1690s produced a framework in which the organic interpretation of fossils was truly functional.

## Minerals and mining

Mineral substances have for millennia been investigated in their own right, and within alchemy, medicine, chemistry and astrology (Goltz, 1972). But the concern to relate minerals to the Earth in which they were found has always been somewhat peripheral. And though there was extensive investigation into minerals in Restoration Britain, there are few decisive trends or landmarks. This is in part because – as Lhwyd recognized – Britain continued to be a backwater in the sciences of minerals and metals, still dependent upon central European expertise (cf. Gunther, 1945: 174).

In Bohemia, Saxony and Hungary, mining and metallurgy were always economically more important than in Britain before the Industrial Revolution explosion of coal-mining (Farrar, 1971). Most mines enterprises in Britain were small-scale, mineralogically and metallurgically more simple than central European operations, and in private hands.

The main endeavour lay in establishing a natural history of minerals. In part this involved programmatically making minerals fit objects for science by stripping them of the supposed magical, alchemical and astrological powers and emblematic symbolism which Renaissance cosmology had accorded them. This paved the way for accumulation of new empirical data, which was one major goal of local natural history. Plot, Leigh, Morton and Robinson each listed the mineral deposits of their respective counties. Christopher Merret's *Pinax* (1667) was an early derivative compendium of Britain's mineral resources (Raven, 1947: ch. xvii). The trend also instantiated the hopes of the Royal Society's 'Articles of Inquiries touching Mines' (*Phil. trans.*, 1666c), and Oldenburg's requests for information. Over seventy articles were published in the *Philosophical transactions* in its first forty years dealing with mineral issues (e.g. E. Browne, 1669; 1670a; 1670b; Glanvill, 1667; 1668). Most of these merely described a specimen, offering some chemical analysis, and a list of properties and uses. Wider intellectual horizons were rare. The wonderful – e.g. combustible pyrites – the useful and the beautiful commanded more attention than synoptic interpretation of the mineral kingdom. And, of course, the great majority of mineral analysis (such as of mineral waters) did not aim at investigation of the Earth as such, and only marginally illuminated its interpretation. But the quantity and quality of knowledge available about local deposits of minerals, metals and useful earths steadily mounted (Gough, 1930: ch. v), being presented to the public either in brief *Philosophical transactions* articles or in general histories. What advance was there beyond this?

The nature and origin of mineral substances were lively issues. Had they existed from the Creation, or were they being continuously generated? Were their generation and properties best explained by alchemical, Aristotelian or mechanical principles? Boyle's work focused on these issues, especially his *An essay about the origine and virtues of gems* (1672). But Boyle wrote as a general natural philosopher, not as one investigating the Earth as such. And most naturalists like Edward Browne were too wedded to empiricism to risk any systematic interpretation of the newly available observations. The most comprehensive work of its time, John Webster's *Metallographa or an history of metals* (1671), was self-confessedly a compilation from earlier Continental writings, whose superiority he freely admitted (1671: Preface).

The development which was ultimately to be of greatest consequence for Earth science was the impulse towards classification of minerals within natural history (Foucault, 1970; St Clair, 1966). It involved a pragmatic drive to distinguish 'genuinely' natural bodies found within the Earth from human artefacts – another aspect of the systematic distancing of man from Nature, within the setting apart of the Earth as a distinct object of scientific investigation.

It also sought to impose an improved, rational, predictive typology of minerals. Gesner, Agricola, Kentmann and Boetius de Boodt had earlier divided the mineral kingdom into earths, stones and minerals, but had concentrated upon describing individual specimens rather than overall taxonomy. As collections grew in size, and the classifying ambition gained momentum, more sophisticated classifications were needed. The most elaborate in Britain was John Woodward's, published as an article entitled 'Fossils' (i.e. mineral objects in general) in John Harris's *Lexicon technicum* (1708–10), and later in its own right as his *Methodica et ad ipsam naturae normam instituta fossilium in classes distributio* (1714), and then emended, translated and dedicated to Newton in 1728 as *Fossils of all kinds digested into a method*. Woodward was aware that many substances were a puzzle to classify. He suggested, for instance, that amber was a natural mineral substance, not a resinous secretion (1695: 217). He debated long-standing boundary problems between human artefacts and natural objects, deciding, for instance, that Cerauniae (i.e. thunderstones) were flint axe-heads rather than stones fallen from the skies (1728: letter v). Despite his strong desire to impose a classification, he recognized that there were no such unambiguous specific differences for distinguishing mineral objects as in the animal and vegetable kingdom: 'nothing *regular*, whatever some may have pretended: nothing *constant* or *certain*' (1695: 171). Being 'no single Character steady, or to be rely'd upon...I am oblig'd to make Use of one or other of them as I see most fit, and conducing to my Purpose' (1728: 22). Woodward developed chemical tests, while continuing to employ external characteristics for identification. His classification finally involved a breakdown into six classes, Earths, Stones, Salts, Bitumens, Minerals and Metals, with subdivisions of the major classes according to tests of sensuous outward characters (touch, taste, composition, colour, etc.). Adopted by J. J. Scheuchzer, it became a basic source of eighteenth-century mineral classification. In itself, the classification was not profound. The salient feature was Woodward's deep conviction of the urgent need for rational and orderly arrangement.

Woodward organized similar material into a different format in his *Attempt towards a natural history of the fossils of England* (1729), listing the locations of the English mineral specimens in his collection. As in all his works, he sought to convey the maximum empirical information while organizing it into a significant explanatory framework. 'Censure is due', he wrote, for those who 'should be perpetually heaping up of natural collections without design of building up a structure of philosophy out of them or advancing some propositions that might turn to the benefit and advantage of the world', for such is 'the true and proper end of collections or observations in natural history and they are of no manner of Use or value without it' (1729: Preface).

The aforesaid work on minerals threw little immediate light, however, on the Earth's structure and history. Investigation of minerals was primarily of specimens involving closet investigation rather than observation *in situ*. So long as minerals and metals were believed generated by unique chemical combinations, stoked by local fires or chemical juices, or activated by special crystallizing tendencies, on the analogy of petrifying waters, mineral studies were not likely to spill over into an interest in structure or strata.

Mining literature *prima facie* might have done just this. But it was still sparse. A number of descriptions appeared in the early

*Philosophical transactions* under the stimulus of the 'Articles of Inquiries touching Mines' and the interests of Sir Robert Moray (Robertson, 1922: ch. viii). Certain Continental authors were translated (notably Alonso Barba, 1674, and Lazarus Ercker, 1683), but no more than a handful of mining books appeared. Furthermore these clustered around issues not immediately concerning the Earth such as mines explosions and safety, and the improvement of engineering and machinery (Moray, 1665a; 1665b). Much of the literature was promotional and less concerned to pioneer a science of ore and seams location than to expound strong patriotic mercantilist reasons for gentry investment in mining enterprises (T. Houghton, 1681; Heton, 1707). I shall briefly discuss in the next section those aspects of mining treatises which were instrumental in the latter development of geology.

### Silent stones

Landforms, fossils and mineral objects had all long been established as subjects of discourse. Grasp of landscape had been needed because man confronted it face-to-face, creating pressing theodicy problems. Fossils were intriguing objects with supposed special powers. Minerals were exploited for their use. Ignored, however, before this period was what was to become the core of 'geology', its outlooks and methods, at the turn of the nineteenth century: the understanding of the continuous structures of rockmass which made up the globe's solid crust. Above all, there was no tradition of study of the strata – their mutual relations, their physical characteristics (faults, outcrops, unconformities, etc.), their order, the relations of lithology to landforms, to fossils, and to the location of minerals and metals.

There were, it is true, strong practical obstacles to such investigation. In most English landscape conditions, the facts of stratification are not immediately obvious. Soil and vegetation cover even the outcrops. Knowledge of lithological structure requires systematic, deliberate and immense industry and travel: it is scarcely picked up casually. The immediate utility of such knowledge was marginal, even for mining purposes. For, being found in veins rather than bedded, most metals could practically be discovered by local trial-and-error methods. Coal was still generally available opencast, or in shallow pits (Nef, 1932). But for good conceptual reasons, also, lithological structure had not appeared worth study. The New Science's goal was to reduce Nature to the minimum number of general laws and elements – without stopping short at fragmentary, local and particular phenomena, as were the strata. Natural history likewise fixed attention upon the eternal, universal verities of the three kingdoms, rather than local diversities. It was widely assumed – not least by working miners – that strata were completely random and jumbled, and the terrain utterly disrupted by faults, caverns, dykes (Woodward, 1695: Preface). With such disorder philosophy had no concern and no purchase. Furthermore, the strata seemed to have no 'meaning', no interest. If, as generally believed, the strata had all been laid down at the Creation, contemporaneously, in much their present arrangement, what significant history could they reveal?[8]

Yet, very gradually, strata became an increasingly significant focus of inquiry. Investigation into landforms, fossils, minerals and general natural history was perforce steadily amassing data upon rocks and their locations, the greatest compilation of this kind being Woodward's *Attempt* (1729). Plot, Leigh, Morton and others were becoming aware, in their routine observations, of the division of regions into limestone, chalk, sandstone or clay areas. Lhwyd and Woodward carefully noted the matrix rocks of fossils. As already noted in the case of John Rawthmell or William Smith, information accumulated within an empirical framework could later be turned to effective use for different purposes. Martin Lister in his plans for a map of England had grasped the regional distribution of rocks. For the first time comparisons were ventured between rocks found in different parts of the country. Lhwyd related the Giant's Causeway to structures on Cader Idris. William Nicolson informed Lhwyd that the Lake District and the north-west of England is '(as you rightly guess) of the same nature with North Wales. Only, I think, we have greater share of metals; and therefore, it may be, fewer formed stones are to be met with in our mountains' (Nichols, 1809: 25–7).

There was also growing recognition that many, if not all, rocks were stratified: that the Earth was – as Plot described it – *'bulbous'* (1677: 56; cf. Challinor, 1971: 59f.). Woodward believed that most rocks were 'naturally disposed into Layers or

Strata; such as our common Sandstone, Marble, Cole, Chalk, all sorts of Earth, Marle, Clay, Sand, Gravel, with some others', and contended that his foreign correspondence had demonstrated this stratification was global in extent (1695: 9). Furthermore, the idea was gaining strength that rocks were *regularly* stratified (V. A. Eyles, 1971: 405), and that identical strata could be continuously traced over long expanses of terrain. A vocabulary for the physical phenomena associated with strata and stratification – dip, strike, fault, vein etc. – was becoming standard (Arkell and Tomkeieff, 1953). George Sinclair defined such a terminology in his *Hydrostaticks* (1672), and it is apparent in mining works like *The compleat collier* (J.C., 1708).

Direct investigation, however, of the structure of the Earth's crust and of the strata as ends in themselves remained uncommon. There was little consciousness that minute investigation of strata, their order and disruptions, the relationships between strata and landforms, fossils and minerals, would order and illuminate those several objects, or the meaning of the Earth itself. Questions about strata structure or history are not central to the *Philosophical transactions* 'Articles of Inquiries touching Mines', or to the natural histories of Plot, Leigh, Aubrey or Robinson.

Nonetheless, attention was gradually fixing on rock-masses. In some cases controversies over other phenomena led to investigation of the strata as a new dimension. For instance, a dispute flared in 1700 between Woodward and, on the other hand, Richardson and Lhwyd over the question whether fossils were found in all strata, and on the tops of the highest hills in England. Woodward claimed that fossils could be found near the summits of Pendle Hill and Ingleborough. Richardson asserted that though he had collected fossils from the limestone *around* Pendle, the higher parts of the hill itself, which were sandstone, yielded none. To prove his point, Richardson then carefully examined the stone and its stratification. He wrote to Lhwyd, 'the stone there is a sand-stone which rises when dug into thin strata about an inch thick: is of a brown colour, and is only made use of for slates to cover houses' (Gunther, 1945: 437). Six months later, in June 1701, Richardson had explored Ingleborough as well, and found the lower slopes limestone but the higher ground of a similarly unfossiliferous grit, and concluded: 'for a confirmation that shells and other marine bodies are not to be found in all rocks (which the Doctor affirms), I will confidently affirm that tho' our country abounds in stone, yet they afford no marine bodies within three or four miles compass from my house, unless the great quantities of *Conchites leviter rugosus compressior* etc in coal-mines, iron-stone and a marble quarry; and this I will make good to the Doctor or any of his confederates, who will take the pains to come hither' (1835: 36).[9] In this way, theoretical disputes which stimulated empirical investigation of terrain almost necessarily led to increased attention on strata and their relations to fossils and landforms.

Moreover, as the science of mining, especially of coal mining, developed, it had to face the issues of strata and lithological structure. Mining surveyors needed to parade to maximum effect their expertise. They had especially to show that their coal-finding art was scientifically predictive; that from examination of relief, or from trial borings, the presence or absence of coal could be deduced from the strata outcropping or superincumbent. Hence, claimed knowledge of the order of the strata, and an ability to make local extrapolations about it, tended to become the stock-in-trade of coal-mining literature (J.C., 1708: 10–17; Heton, 1707: ch. xi; Porter, 1973b: 336f.). However dubious much of his 'expertise', the logic of a mines expert's situation ultimately forced him to address himself to relations of strata to the Earth at large. And, of course, the operations of digging shafts perpendicularly through the strata, and extracting minerals laterally along them, also gave immediate practical experience of the geometrical parameters of stratification.

Lastly, two contemporary theories, Hooke's and Woodward's, made strata central to their views of the nature and history of the Earth.[10] For Hooke strata constituting present land-masses had been formed slowly by natural processes on the sea-bed, with organic remains becoming embedded in them, and subsequently raised by earthquakes. Strata were thus a witness of the changes the Earth and life had undergone. For Woodward, the strata settled after the Deluge in order of their specific gravities imprisoning organic remains, as could be explained and predicted by Newtonian philosophy. This order proved the Deluge, and could be put to empirical test. Woodward's work in particular earmarked strata for all future theories of the Earth, and invited attempts to test his strata hypothesis empirically.[11]

Systematic observation of the Earth in Britain was undertaken for the first time in the latter half of the seventeenth century. The first basic task was to compile a complete inventory of natural objects. This endeavour was fruitful in establishing a momentum which was to last through the eighteenth century. It gave rise to more sophisticated aspirations and techniques in seeking empirical solutions to specific problems, such as the determinants of land-forms and the nature of fossils. These problems were resolved in an ad hoc manner. However, this was not simply through the application of new data, but through new investments in high-level philosophy of Nature which offered a fresh interpretation of such data. For the very rendering of the Earth as an object for science was undergoing constant modification, partly under pressure from the scientific movement at the most general level, partly because man's actual relationship to the Earth was changing, and partly as a precondition for the activity of field inquiry.

Different features of the Earth – landforms, fossils, minerals – still continued to be studied within their own distinct intellectual traditions. However, local natural history was beginning to integrate investigation of these objects within the new practical and conceptual framework of an organized physical geography. Furthermore, empirical study was breeding practices and view-points which were formative for the possibility of geology. Under the stimulus of natural theology and the New Science, investigation of landforms began to subsume terrestrial stability and change within uniform natural laws. The Earth began to be studied as a regular economy. Understanding of fossils took on historical dimensions. And confidence in order and uniformity became the keynotes of a study of rock deposits and strata. These were inchoate tendencies away from an ideal, universalizing natural history towards consideration of the globe as an organized system of forces and products.

# 3

## The re-creation of the Earth

### *Historical contexts*

Many theories of the Earth – and far more replies – were produced in the generation straddling the end of the seventeenth century. Early nineteenth-century geologists dismissed these as bare theological speculation, and historians of geology have generally endorsed this judgment: 'these early fables of geological science should be read by all who are in need of mental recreation and who possess the required leisure and a certain sense of humor' (Adams, 1954: 210).[1] This chapter attempts to establish a framework for understanding late seventeenth-century theories, in themselves, and within the development of Earth science.

Firstly, however, it is worth stressing their importance. For these theories were the work of the leading scientific intelligentsia. They included such first-rank natural philosophers as Hooke, Halley, Ray, Keill, Whiston, as well as the age's foremost students of the Earth, like Woodward. Newton first admired Burnet's theory ('Of our present sea, rocks, mountains, etc., I think you have given the most plausible account'), then formulated his own ideas, and finally gave his approval to Whiston's (Turnbull, 1959, ii: 329; Whiston, 1698: Preface). John Locke wrote of Whiston's theory, 'I have not heard anyone of my acquaintance speak of it, but with great Commendation, as I think it deserves' (Davies, 1969: 86). Other virtuosi, such as Samuel Pepys, Addison and Steele, gave their attention to Burnet (Davies, 1969: 72).

If these theories failed to resolve the problems of the Earth, it was not through stupidity, but because of intrinsic difficulties, and the contradictory demands made by their philosophies of Nature. The theorists generated much satire; but that proves public fascination, and perhaps the intellectual daring of the theories, not incompetence. British theories were avidly read and translated on the Continent during the eighteenth century, and naturalists such as J. J. Scheuchzer abandoned old ideas on discovering them (Zittel, 1901: 20f.; V. A. Eyles, in Schneer, 1969). Likewise, throughout eighteenth-century Britain, debate was conducted explicitly around the theories and outlooks of Burnet, Woodward and Whiston. Though dissociating themselves from them, Buffon, Raspe, Whitehurst and others nevertheless believed that *these* were the theories to be refuted in order to clear the decks.

Equally, to cast these works aside on the grounds that they were *theories* in an age supposedly of Baconian empiricism and 'hypotheses non fingo' is totally to misread both the philosophy and the practice of science in that age (cf. Buchdahl, 1969; Butts and Davis, 1970) – not least, because much theorizing about the Earth itself had a decidedly tentative tone. Burnet averred that thought was free, and that his theory was a mere hypothesis: 'there is no Chase so pleasant, methinks, as to drive a Thought, by good conduct, from one end of the World to the other' (1684: 6). Hooke argued the empirical basis of his theories, which he defended as heuristic (Oldroyd, 1972). For all his personal arrogance, Woodward offered his theories as *Essays* and *Attempts*, and stressed 'From a long train of Experience, the World is at length convinc'd that Observations are the only sure Grounds whereon to build a lasting and substantial Philosophy' (1695: 1).

The most common dismissal of these theories, however, is that they were not the outcome of empirical inquiry, but were rather amalgams of imagination and speculation, *a priori* metaphysical and natural philosophical notions, religious Revelation and uncritically regurgitated ancient wisdom, derived from books, sanctified by age and authority. This is partly true – though only partially. But it is generally stated in a form which begs the historical question. It raises three issues. How empirically based were these theories? That is, to what extent did they grow out of the natural history inquiry orchestrated by the founding of the Royal Society? This question will be discussed later.

Secondly, how should we evaluate these theories, given that they were built with many other forms of knowledge in addition to direct empirical inquiry? For all late seventeenth-century men of science, whatever their genius or specialist discipline, believed in the multiplicity of modes and sources of inquiry after truth (Manuel, 1963; McGuire and Rattansi, 1966). Theorists of the Earth were not unusual in this respect. I shall explore below the different kinds of evidence and argument taken as valid for a theory of the Earth, and how fundamental religious and philosophical beliefs determined the form of theory which contemporaries could realize.

The third question is the simplest to resolve: were not these theories *merely* exegeses of Genesis, tailoring nature to the Scriptures, either by choice, or under pressure from public and ecclesiastical opinion? But this does not seem true. I know no evidence that any theorist – except perhaps Halley – significantly felt inhibited by any such pressure. They wrote what they believed.[2] Furthermore, they themselves discriminated between merely exegetical philosophy and empirical inquiry, and many actually chose to structure their work into separate exegeses of Scripture, philosophical discourse, and consideration of empirical evidence. Thus Thomas Burnet argued the first eight chapters of the *Theory of the earth* from rational and philosophical necessity, and then announced his intention to corroborate these points from 'Effects', in ch. ix, headed, 'The Second Part of this Discourse, proving the same Theory from the Effects and present Form of the Earth', believing that 'Truth cannot be an enemy to Truth, God is not divided against Himself' (1684: 3).

To see these theories as obediently toeing the theologically orthodox line misses the historical point. For such theological commitments were *themselves* in the melting-pot. Late seventeenth-century intellectuals had been inducted into a culture rich in theological and philosophical wisdom, natural lore, myths about man's origin and development, prophecies about the future of man and the world, allegories, symbolism, occult knowledge, mysteries. Because it was a culture overbrimming with such heterogeneous claims to truth, the critical methods, philology, sceptical temper, and sheer mushrooming of new facts of the Age of Restoration and the early Enlightenment, were shaking its foundation (Hazard, 1964; Manuel, 1959; Kubrin, 1968). Empirical natural history was espoused to resolve the narrower ambiguities of the meaning of the Earth – problems of fossils and minerals. So, in the same way, theorists of the Earth took apart the world of traditional cosmogonical discourse, and sought to expose its flaws, resolve its ambiguities, and reconstruct it on a fit basis for a new age.

This rarely entailed anything so mechanical as putting traditional wisdom to empirical test, finding it wanting and simply discarding it, though this occasionally happened, Ray and Woodward both being scathing for instance about the 'wisdom' of the Ancients (Woodward, 1695: 56; Lankester, 1848: 229f.). Rather, empirical evidence was weighed in order to decipher the 'real' meaning of traditional wisdom. The process was one of revaluing, rather than rejecting, the components of knowledge in the world of traditional discourse. Thus, Burnet used natural necessity and the examination of effects to confirm that Genesis was not a literal account of the Creation but an allegory for the 'vulgar' (1684: Review). The evidence of strata and fossils convinced Woodward that the Deluge must have been effected by immediate miraculous agency. Investigation persuaded Hooke that Ovid's *Metamorphoses* contained a deep philosophical understanding of the Earth. It would be unhistorical of course to want to imply that these theorists open-mindedly put problems of interpretation of Scripture, traditional wisdom and philosophy to the test of Nature. Clearly, theorists largely found in Nature what they hoped to discover. (Yet not always. For example, in his *Miscellaneous discourses* Ray frequently admitted that the empirical evidence was inconclusive.) The real point, however, is that empirical natural evidence was now becoming the approved testing ground, ostensibly at least, for rival interpretations of Earth history, the destiny of mankind, the order of Nature, the mode of God's activity in the world – in short, of cosmologies and cosmogonies. Perhaps for the first time, certain fundamental beliefs were being held up for empirical confirmation at the tribunal of Nature. Of course at this stage most theories were reinterpreting, modifying and broadly reinforcing traditional cosmogony, rather than challenging it. Yet the way was opened, once the ground of debate had shifted to natural evidence, to such confutation in the future.

And even then, quite radical pressures were creating stresses on the traditional cosmogony. Burnet did not merely find (as the popular jingle ran) that 'All the Books of Moses/Were nothing but supposes', but allegorized the Fall, and seemingly declared that in the physical world God worked exclusively by second causes (Redwood, 1976: 110f.). Hooke and Aubrey thought 'the World is much older than is commonly supposed' (Aubrey, 1685: 112; Hunter, 1975: 59). Hooke accepted extinction. Beaumont hinted at the eternity of the Earth (Davies, 1969: 11–12). Furthermore, controversy revealed disagreement on all major issues, from microscopic facts up to total philosophies of Nature. In sum, these theories were not exercises in Biblical exegesis, but rather many-sided projects of the mind attempting to bring order out of cosmogonical chaos.

While differing in detail, late seventeenth-century theories shared certain fundamentals. All conceptualized the Earth in a context of Divine Creation, purpose, order and sustenance. All believed the Earth had been created for man, though not necessarily exclusively for him. It was, consequently, to be understood teleologically and anthropocentrically. This was the outlook of Hebraic Christianity, and also of Hellenistic philosophy and Classical culture; of rational religion and natural philosophy. Though repudiated in the strategy of physical science, teleology lost no ground in the conceptualization of the general economy of Nature until long after the sceptical attacks of Hume and Kant (Glacken, 1967; Humboldt, 1845–62). Late eighteenth-century Deistic geologists such as James Hutton and Erasmus Darwin needed to believe in the rational, man-centred purposiveness of the Earth no less than earlier Protestant Christians.

Earth history, therefore, hinged on human history. Habitat was made for inhabitant. Such a view justified another general tenet, that the Earth had not existed for many thousands of years, and quite possibly would not endure much longer.[3] Within a cosmology in which an eternal God had created Time and the world simultaneously, a long time-scale was no more plausible than a short one. Perhaps the reverse, for evidence suggested that man had not lived for long on this planet, as the Bible, taken as God's revealed Word and as one of the oldest – perhaps the oldest – relics of civil history, affirmed. "Tis the Sacred Writings of Scripture that are the best monuments of Antiquity' (Burnet, 1684: 4). The Bible was corroborated by the chronologies of all nations with reliable records, and reaffirmed by natural evidence. Man had not yet had time to populate – or repopulate – great expanses of the globe. Civilization was still in its infancy. Man had not subdued Nature. The earliest records of all civilizations contained memorials of the creation and youth of the Earth (Stillingfleet, 1662; Hale, 1677). Surely folk memory and recorded evidence were good testimony that man and the Earth were almost coeval. 'Time we may comprehend', wrote Sir Thomas Browne; 'it is but five days elder than ourselves'. Why should the Earth have long pre-dated man, untenanted by the one rational creature formed in God's image to conserve and understand it?

Within such fundamental background beliefs many grave tensions remained concerning the workings of Earth history – the mechanism of Creation and the Deluge, and the laws maintaining the global economy. All such questions, as well as raising physical and technical issues, possessed overt teleological dimensions relating to human perceptions of Nature and the problem of theodicy.

Such questions churned up conflicts within basic beliefs. The most critical was of the mode of God's providential operation, the higher rationality whereby God intervened miraculously as well as governing the world by natural laws. Facts, reason and Scripture supported all positions. Miraculous interventions proved that God was active and personal, but might suggest faults (or, at least, narrowness of vision) in the original design, and threatened to put an end to philosophy. Action through laws demonstrated God's wisdom but scarcely His freedom, love or wrath. Such issues were wide open for debate.[4]

In recreating the nature and history of the Earth, what kind of knowledge had to be accommodated and integrated? Accurate and well-attended facts were one agreed legitimate form of knowledge. In the next section I shall explore the use in these theories of the data built up in the previous generation. Secondly, natural theology, general philosophies of Nature, and laws from other sciences, had to be harmonized. Though certain metaphysical systems were *non grata* in late seventeenth-century England, postulates of the kind attached by Thomas Robinson to his *Natural history of Westmorland and Cumberland* (1709) were organizing principles for any theorizing of the Earth:

*Such* 'APHORISMS, DEFINITIONS and AXIOMS' *as we have made use of in this Treatise*

*Of Nature*

Nature's Productions are never in vain.
Nature's Productions are not by blind Chance.
Nature never works in haste.
Nature's Productions are not by precipitous Leaps, but by gradual Motions.
Nature never does that by Much that may be done by Less. . .(p. 114)

Nobody challenged in principle the use of rational deductions from necessary attributes of God or Nature (for example, the economy of Nature) – though some challenged particular deductions in particular cases.[5] Indeed, many theorists formally set out their own comprehensive philosophies of Nature, as Nehemiah Grew's *Cosmologia sacra* (1701), John Ray's *The wisdom of God* (1691) and John Keill's *An introduction to natural philosophy* (1720).

Thirdly, within the anthropocentric geocosm, human testimony and ancient wisdom weighed heavily as knowledge, being the savings bank of man's sagacity, and man's only eye-witness of early history (D. P. Walker, 1972; D. C. Allen, 1949). Furthermore, the Ancients were supposed wiser than the Moderns, having lived when intellect was undecayed, and knowledge less corrupted by transmission, and perhaps having received special revelation from God.

Fourthly, Scripture was knowledge, partly as a special case of Ancient Wisdom, and partly because Revealed. With the opaque corpus of Ancient Wisdom in general, but most acutely with Scripture, hermeneutic problems were labyrinthine and controversial. Could Biblical passages really be irreconcilable? Burnet for example treated the Genesis cosmopoeia as unphilosophical, but embraced the wisdom of the Book of Job, the Psalms and St Peter. The two potential extremes of interpretation – that the Bible was, with regard to the physical world, merely erroneous or fabulous, or, conversely, that it offered immediate, full and literal truth – found no favour amongst an elite accustomed to delicate and judicious classification of reality. Those like Burnet with neo-Platonic leanings tended to take the Bible rationally and philosophically. Those with stronger fundamentalist Protestant leanings, like William Whiston, favoured variants of literalism. Thus while Burnet pointed up the perils – ''Tis a dangerous thing to

ingage the authority of Scripture in disputes about the Natural
World, in opposition to Reason' (1684: Preface) – Whiston re-
torted that Burnet contradicted the letter of Scripture (1696: 77f.),
and went on in his 'Discourse concerning the Nature, Stile, and
Extent of the Mosaick History of the Creation' to establish

*POSTULATA*

I. The Obvious or Literal Sense of Scripture is the True and Real one
where no evident Reason can be given to the contrary.

II. That which is clearly accountable in a natural way is not without
reason to be ascrib'd to a Miraculous Power.

III. What Ancient Tradition asserts of the constitution of Nature or of
the Origin and Primitive States of the World, is to be allow'd for
True, where 'tis fully agreeable to Scripture, Reason and Philosophy.
(Whiston, 1696: 95)

All admitted, however, in their different ways that the matter and
style of Scripture had been accommodated to the understanding
of the early Jews, and that true physical meaning needed to be
winnowed from the chaff.

The age fervently sought unity of knowledge, for unity promised
stability. In particular, natural and evidential theology, both
of fundamental significance within late Stuart vindications of
Christianity, specifically encouraged integration of natural and
scriptural knowledge (Cragg, 1964; Pattison, 1859). Bacon had
earlier urged that one could not 'search too well in the Book of
God's Word or the Book of God's Works, Divinity or Philosophy'
– adding only the proviso that one should not 'unwisely mingle
and confound these distinct learnings' (1605, i: 3). The great late-
century natural theologies, such as Ray's *The wisdom of God*
(1691) and Grew's *Cosmologia sacra* (1701), were less cautious
about mingling and confounding. Boyle himself had specifically
urged that natural philosophers should become natural and
evidential theologians, corroborating Genesis:

And indeed so far is God from being unwilling that we should pry into
His works, that by diverse dispensations He imposes on us little less than
a necessity of studying them. For first He begins the Book of Scripture with
the description of the Book of Nature. We may next observe that God had
some knowledge of His created Book both conducive to the belief and
necessary to the understanding of His written one. (Boyle, 1774, ii: 19)

Within this powerful tradition, theorists claimed that Nature
collaborated the Mosaic accounts of the Creation and the Deluge.

Woodward wrote that as a result of 'freely' comparing the Mosaic account and natural evidence he had proved 'the Fidelity and Exactness of the Mosaick Narrative of the Creation' (1695: Preface). For Whiston,

> Whatever incompetent Judges may say, nothing will so much tend to the vindication and honour of reveal'd Religion as free enquiries into, and a solid acquaintance with, (not ingenious and precarious Hypotheses but) true and demonstrable principles of Philosophy, with the History of Nature, and with such ancient Traditions as in all probability were deriv'd from *Noah* and by him from the more Ancient Fathers of the World. (1696: 63)

Similarly, Burnet claimed that sacred history had been 'confirmed anew' 'by another Light, that of Nature and Philosophy', and John Ray hoped that he could 'confirm the Truth of the History of the Deluge'.[6] All these traditions of knowledge had to be integrated into a total theory which would respect the internal logic and criteria of each: one which would embody a usable cosmic myth while also satisfying the standards of objective scientificity upon which the credit of the project now depended.

### The structure of theories of the Earth

The cosmogonies of the late seventeenth century were attempting to project a *myth* of the Earth, in the sense of a total, self-justifying framework, which would weave all knowledge together, and explain each partial facet – whether the role of volcanoes or the diaspora – in respect of its position within the whole. But significantly – and here they are unlike, say, *Paradise lost* – these were *scientific* myths, or *realizations*, which aimed to have the strength of being rational and empirical achievements of objective truth, hence also of being open to debate and correction.

Though differing in myriad details, certain constellations of structures appeared.[7] One characteristic way of seeing the Earth was that which informed Thomas Robinson's *The anatomy of the earth* (1694) and William Hobbs's *The earth generated and anatomized* (1715). The Earth was taken as a living body, or animal, and explained in terms of its structures and functions within a fully articulated macrocosm–microcosm correspondence theory. Anatomy and physiology provided the teleological and functional explanatory schema. Theories of this kind were rooted in the Hermetic, alchemical and mystical currents associated with Paracelsus, Fludd, Gabriel Plattes and Kircher. But by the end of the seventeenth century they were ignored or rejected, and un-influential in the further growth of the science.[8]

The dominant way of realizing the Earth came to be one in which significance hinged upon logical and chronological sequence. Most theories – those for example of Hooke, Burnet, Ray, Woodward, Whiston, Keill, Arbuthnot – presupposed that understanding the Earth entailed a historical account of its origin and development, relating its present nature to past events, and perhaps to its future state. Confidence in such *historical* explana-tions followed from certain Christian, and especially Protestant, modes of thought, which attributed the significance of particular events to their place in a providential sequential chronology deter-mined by God.[9] Theorists aimed to find 'the Cause and Reason of the present Figure, Shape, and Constitution of the Surface of this Body of the Earth, whether Sea or Land, as we now find it presented unto us, under various and very irregular Forms and Fashions and constituted of very differing Substances' (Hooke, 1705: 334).[10]

They could agree the Earth had not eternally been in its present condition. The testimony of disputed rocks proved that past changes had occurred. Human evidence told of change. In any case, present processes, however slight, were modifying the Earth, however slowly. Hence no theory sought to 'explain' the Earth by trying to prove that it had always existed in its current state. The problem, in other words, was to explain change.

Fundamental, then, was whether past changes could fully be assimilated in terms of those with which contemporaries were familiar. Two theorists staked claims for the size and scope of present change: Burnet and Hooke (and, probably derivatively from Hooke, Aubrey). It is no accident that they were among the most senior theorists. Burnet in particular was much influenced by Cartesianism, and especially by Descartes' emphasis in his *Prin-cipia philosophiae* (1644) upon the development of the Earth by natural regular causation (Farrell, 1973: ch. ii). They grew up at a time when the decaying Earth still compelled assent. Yet Burnet did not believe that current decay, due to weather, sea and rivers, could account for what *he* conceived the evidence of past terrestrial change – the original separation of land and sea, mountains and valleys, highland and lowland. His championing of the potency of current processes rather served to prove that the Earth could not be eternal – in fact, could not be immensely old.[11]

For Hooke, however, the power of known present causes was the precise scientific guarantee of the plausibility of his philosophy of past revolutions. He re-echoed Ovid's view that 'All things almost circulate and have their vicissitudes' (1705: 313). He believed in the Earth's decay, being impressed by evidence for volcanoes, earthquakes, new islands, subterranean caverns, landslips, tidal waves and denudaton. His commitment to the New Science led him to seek explanations within known processes of Nature, and he shunned miraculous or extraordinary causation. Being empirically and metaphysically convinced that figured stones could not be other than organic relics petrified while entombed within sediment on the sea-bed, he looked to the land-consolidating and uplifting powers of earthquakes and volcanoes as the only credible explanations of revolutions of land and sea (1705: 295f.). Hooke, however, believed that though such past forces were of the same kind as existing causes, they had diminished as the Earth approached senescence.

Burnet and Hooke could place change at the centre of the stage, because decay articulated their theodicy. Their bid for scientificity depended utterly upon the rationality of Nature. Later theorists, however, possessed deep intellectual and emotional investments in the sovereignty of God and the order and stability of Nature, and they used their radical scepticism to bludgeon hypothetical, unsubstantiated notions of change and to sanctify Nature's *status quo*. By the 1690s Newtonianism for similar reasons was also lending its authority to a highly passive view of the globe.[12] Newton's vision of the solar system – indeed, the whole universe – as a stable system of forces was readily applied by analogy to the terrestrial economy. His conception of matter as hard, inert and impenetrable, not essentially possessed of force, motion or life, was explicitly used to guarantee a passive Earth, especially in his protégé William Whiston's geometrically arranged *New theory of the earth* (1696: Bk i, Lemmata). Newton's matter theory and Whiston's theory of the Earth express the same intellectual convictions. Whiston believed that to see matter or Nature as endowed with

life or 'tendencies' was a vulgar, unphilosophical error. He feared
that concessions to any kind of hylozoism, hylarchic principles or
to Cartesianism, must lead to atheism.

Theorists like Woodward, Whiston and Arbuthnot of course
accepted that the Earth had undergone changes.[18] Their causes,
however, could be seen as not inherent in Nature itself, but dependent upon God's immediate intervening will. Nature's passivity
proved God's Omnipotence. Whiston's 'The *Idea* and Nature
of God includes *Active Power*. . .the *Idea* and Nature of matter
supposes *Intire Inactivity*' (1696: 222) crystallized the intimate
intellectual links of Protestantism and Newtonian metaphysics in
contemporary theories of the Earth (Hooykaas, 1972). In response
to the Cartesian sufficiency of Nature ('Give me extension and
movement and I will remake the world'), Ray accused Descartes
of trying 'to solve all the *Phoenomena* of Nature, and to give an
account of the Production and Efformation of the Universe, and
all corporeal Beings therein, both celestial and terrestrial, as well
animate as inanimate, not excluding Animals themselves, by a
slight *Hypothesis* of Matter so and so divided and mov'd' (1701:
44). He reasserted his faith in the active finger of God, to pinpoint
his programmatic conclusion that mere natural forces were not
competent to account for the phenomena.

Thus, most theorists believed that such forces as volcanoes,
earthquakes, the effects of climate and of sea on coastlines, were
not sufficiently universal or powerful to solve the chief explicanda
– fossils, strata, the distinction between land and sea, mountains
and valleys. In taking this stance, they sought to vindicate both
Design and the rigour of their own investigation. Hence there
must have been periods of extraordinary forces operating during
Earth history. This is not to say that all fell back upon miracles,
for some theorists such as Burnet insisted that extraordinary
events, such as the Deluge, nevertheless lay within the regular
course of natural causation. But all sought to explain Earth history
in terms of the two great formative epochs, related to them in the
Bible, in the mythology of the East, and in much Classical literature: the Creation and the Deluge.

Theorists were adamant that empirical evidence, philosophy
and religion proved that the Earth could not have existed from
eternity, and that it could not be self-creating or self-sustaining (e.g.

Stillingfleet, 1662). The Earth itself had been created. Further-more, no known, natural process was creating massive areas of consolidated land, was differentiating land from sea. Most agreed that no current processes were raising up high mountains, pro-ducing fossils deep within strata, sculpting valleys, shaping the coast-lines of entire continents. Practically all theorists, in other words, attributed such events to some special period of creative activity in the youth of the Earth, during which either by divine fiat or by laws not operative subsequently the Earth had emerged from Chaos into a form somewhat akin to its present one.

But they were also aware that one such creative revolution alone could not explain the Earth's physical face. If Creation left the Earth a smooth paradise without seas (as Burnet believed), how had inequalities of land, and the boundary between land and sea, arisen? If strata had been formed at the Creation before the origin of life, how had organic fossils become deeply embedded in the rocks? If the strata had originally been created in a horizontal condition, how had they come to be raised and lowered, faulted, disrupted, displaced?

Once again, theorists' realizations of Earth history made actual processes seem incompetent to account for these changes. Seeking the origin of petrifactions, John Woodward methodically ex-amined all hypothetical natural causes: that they were 'kitchen refuse'; that they had become lodged in rocks via subterranean channels leading from the sea; that they were debris of local floods, tidal waves, or the original covering of the land by the sea, or had been left by supposed retreats of the sea. He eliminated all of these because 'scarce any of all these alledged, had the least countenance either from the present face of the earth, or any credible and authentick Records of the ancient state of it' (1695: 40). Thus, fossils could only be the product of a unique and miraculous Deluge.[14]

Hence most theorists attributed the great revolutions of the Earth subsequent to the Creation to the Noachian Deluge. This explanation had much plausibility. Deluges on a smaller scale were well-attested events. Most cultural traditions included major widespread deluges. A universal Deluge could account for global phenomena such as fossils and strata disruption. Furthermore, deluges were likely instruments for the kind of changes in question –

especially gouging out deep valleys, shifting coast-lines, and fossils strewn over the land. Of course, each theorist had his own view of the Deluge, interpreting Genesis differently, and offering different mechanisms. Woodward and Keill invoked a miracle; Whiston, the approach of a comet; Halley, a shift in the Earth's centre of gravity; Burnet, a contracting of the Earth's crust. Likewise, they attributed different effects to the Deluge. Woodward believed that the Deluge had taken apart the Earth's total fabric, which then needed to be reformed in a second Creation. Ray thought the Deluge's effects essentially superficial. Interpretation of the Deluge varied according to what it needed to explain.

Most theorists agreed that, since the Deluge, Nature had operated according to familiar principles and actual causes. For Burnet this meant continual decay, but for later theorists it entailed a regular and well designed economy which maintained the habitable Earth in stability. Whiston described the present system of Nature as 'regular, beautiful, permanent' (1696: 286). Woodward saw the world existing in 'the most excellent and beautiful order' (1695: 60). Providence's intention was that present causes should preserve, not destroy, the world. Whereas Burnet had believed natural forces would in time destroy the Earth, Ray, Woodward and Whiston all argued that the Earth would maintain its present order until God decreed.

John Woodward's *Essay* (1695) epitomized this pattern of Earth history. He conceived the Earth's economy as a stable, ordered and benevolent system, with minimal forces producing minimal change. The ante-diluvial Earth 'was not much different from *this* we now inhabit' (1695: 248). It 'hath been in much the same condition that it is at this day, ever since the time of the Deluge' (47); and 'the Earth, Sea and all natural things will continue in the state wherein they now are, without the least Senescence or Decay, without jarring, disorder or invasion of one another' (61). He explained the presence of fossils in rocks by invoking a universal Deluge, during which all rocks disintegrated as a result of loss of gravitational cohesion, the reliquiae of creatures were introduced, and the rocks resettled much as before, in order of specific gravity. The Deluge was immediately occasioned by a sudden miraculous influx of water 'wherever it may be now hid' (163), for 'the Deluge did not happen from an accidental Concourse of

Natural Causes, as the Author above-cited [Burnet] is of Opinion. That very many things were then certainly done which never possibly could have been done without the Assistance of a *Supernatural Power*' (1695: 165). For Woodward, the Deluge explained almost the whole range of terrestrial problems, in context of man's physical and moral condition (cf. Lyell, i, 1830: 37). Its evidences were daily proof of 'the Fidelity and Exactness of the Mosaic Narrative of the Creation'. Since this supernatural event, the history of the Earth had been a history of quiescence (93).

Strategically, these theories presented a most apt schematization of Earth history. They impressed it with defined order, by dividing it into a sequence of significant states (before the Creation of life; the Creation of life, and of man; the Deluge; post-Deluge), separated by great revolutionary events of human significance – Creation, Deluge, Conflagration. Burnet thus declared that he would only deal with such 'great revolutions of the globe', as did the Bible, for these were 'truly the hinges upon which the Providence of this world moves' (1684: Preface).

John Ray's *Miscellaneous discourses* likewise illustrate the imposition of ordered purpose. Ray wrote the *Discourses* in a hurry, partly out of old notes (1692: Preface). They consist of a major discourse, showing that no natural terrestrial forces appear to be advancing that general Conflagration prophesied by St Peter which will give the earth its quietus; broken up by two long digressions, the first on the Deluge, the second on the Creation – hence the apt title 'Miscellaneous'. Revising for a second edition, however, Ray systematized them into a full history of the Earth, entitled *Three physico-theological discourses*, the first now on the Creation, the second on the Deluge, and the last on the Conflagration (1693: 50f.). The full title of Whiston's theory equally reveals the same ordered structure of Earth history: *A new theory of the earth, from its original to the Consummation of all things wherein the creation of the world in six days, the universal deluge, the general conflagration as laid down in the Scriptures are shown to be perfectly agreeable to reason and philosophy* (1696). The theorists succeeded in conceptualizing the Earth as an object, distinct from man and possessing its own laws, while at the same time retaining a powerful notion of it as a habitat, subject, by God's will, to human needs and purposes.

Furthermore, this theoretical structure had the attractive quality of orchestrating – rather than challenging, as *geology* did later – contemporary educated common sense. The Earth's history, thus interpreted, confirmed the Biblical and rational perception that rocks antedated life; that sea creatures had preceded land animals; that plants and animals antedated man. Englishmen crossing the Alps could fully appreciate Burnet's view that such mountains had not been created by everyday processes.

Lastly, these theories absorbed within themselves the concepts of contemporary physical science. Newton's *Principia* had solved the celestial system: the terrestrial system now demanded its Newton. Whiston, Newton's protégé, significantly set out his *New theory* (1696) rather in the form of Newton's *Principia*. John Hutchinson revealingly called his counterblast *Moses's Principia* (1724–7). Most of these theories strove to integrate Newtonian physico-chemical mechanisms to explain the great events of the world. Newtonianism offered theories prestigious quantification, and validated them within the range of simple universal laws of the physical world (cf. Kubrin, 1967).

Woodward's whole theory hung on Newtonian gravity, which, like Newton, he insisted was not essentially inherent in bodies, but was 'Intirely owing to the direct Concourse of the Power of the Author of Nature, immediately in his hands, . . .the prime hinge whereon the whole frame of Nature moves'. Gravity produced the cohesion of the original chaotic particles to form distinctive rocks, whose specific gravities precisely determined their position in the order of strata (1695: 51, 75). Furthermore, for Woodward, the Deluge was occasioned by *suspension* of the laws of gravity. Burnet, Hooke and Halley used gravity in a different way, showing that the Deluge had been caused, or at least accompanied, by shifting of the Earth's centre of gravity – a conception which appealed to Ray, and to Whiston before Newton himself voiced his disapproval.[15] Explanations of the Deluge through the mechanism of land collapse derived some of their popularity from the central role they gave to gravity. Newton's own suggestions for the Deluge hinged upon it (Turnbull, ii, 1959: 329).

The scheme most strikingly derived from natural philosophy was Whiston's *New theory*. This was set out *modo geometrico*, as a series of deductions from Newton's metaphysics of matter, his laws of motion, and his physical astronomy (1696: Lemmata). For Whiston, as for Newton, the Genesis account of Creation out of Chaos referred only to the Earth. Its oblate spheroid figure, as proved by Newton, demonstrated its original chaotic state (1696: 53, 72). The Earth's transition from a chaotic to an ordered condition had a cosmic rather than a strictly terrestrial explanation (viz., the Earth was once a comet). So stable was Whiston's view of the terrestrial economy that subsequent changes, the Deluge and Conflagration, had to be effected by external means, by comets passing close to the earth (1696: 73, 437).

Certainly, there were alternative general physical explanations, such as Burnet's mixture of Aristotelian and Cartesian physics, or Robinson's neo-Platonic organicism. A few, such as Ray – significantly, chiefly a natural historian – held back from all-embracing causal physical accounts of Creation and Parousia. But by 1710, the Earth had fallen firmly within the attractive field of Newtonian explanatory models.

That so many competing and contradictory theories vied against each other is no mark of incompetence. Each strategically focused upon a few key problems without being able to weave all other outstanding issues into an overall fabric with the power to convince others. Newton had his principate, but of Earth science there was no king. Yet if no individual theory triumphed, the broad structure of the realization of Earth history was becoming well-established for the future.

### Boundaries and bridges

The seventeenth century saw no unification of Earth science. A tradition of geocosmic physics flourished, expressing itself in the genre of the theory of the Earth (Roger, 1973–4). Similarly a natural history practice steadily gathered strength, hingeing on Baconian fieldwork and issuing in the *catalogue raisonée* of terrestrial objects. But these traditions strongly fortified their boundaries against each other, and there was generally little common ground between them. At this time most theorists of the Earth were not – nor were seeking to be – fieldworkers. Burnet, Hooke, Whiston, Warren, Croft, Halley, Keill had not in any real sense personally practised fieldwork. For Burnet, the certainties of Cartesian mechanism, for Whiston the truths of Newtonian celestial physics, transcended the uncertain world of empirical evidence from the ruins of the Earth. *Per contra*, fieldworking naturalists like Lhwyd set their face against all contemporary forms of theorizing. Those who sought to realize an interpretation of the Earth which integrated fieldwork and theory, a concern for objects, interpreted *within* an ambition to schematize total Earth history, were the exception: Woodward, Ray and Hooke perhaps come to mind.

But to speak of the resilient independence of isolated traditions of scientific practice is not to deny that there were levels at which those diverse practices deeply interacted – and thereby produced the conditions for their own transcending. For there was important interplay between fieldwork and theorizing. The theories of the 1690s would hardly have been possible without the generation of intense fieldworking launched in the 1660s. In Ray's and in Beaumont's case, early fieldwork produced a later readiness to generalize, deploying the mounds of empirical research published in the interim.

Of course, other kinds of interplay between theories and observation grew up. Fieldwork came to be undertaken to confirm or challenge particular theories. Morton investigated Northamptonshire in the light of Woodward's belief that strata lay in order of specific gravity. Hauksbee used vertical sections of strata to test the same hypothesis (1712). Halley proposed experiments on the salinity of seas to determine the Earth's age (1715). Theories could themselves open up observation programmes. And of course, controversies around theories deployed data in the cut-and-thrust of assertion and denial of fidelity to nature.

Under pressure from specific criticisms and new evidence, the succession of theories demonstrates tangible intellectual progression. It is in part a trivial but necessary story of error being recognized and corrected. For example, Burnet still believed with the Ancients that the Earth was egg-shaped, but Newton subsequently convinced theorists it was rather an oblate spheroid (cf. Whitehurst, 1778: 2). But more important was a broadening consciousness of important criteria. In 1681 Burnet could publish his *Telluris theoria sacra* showing no apparent awareness of stratification, or that fossil evidence might bear in some way upon his thesis. Burnet's controversy with Warren (1690; 1691) brought the fossil issue out into the open. Woodward's *Essay* (1695) irreversibly required that all future theories address themselves to the facts of stratification. But at a more fine-textured level, Woodward himself had not 'read' various problems which could be posed by fossils and strata – e.g. the meaning of the fact that most known fossils were exuviae of *sea*-creatures – for Woodward's Newtonianism led him to differentiate fossils according to specific gravity not zoological type. Shortly afterwards, however, Abraham de la Pryme grasped this as a problem, and argued on the strength of it that land and sea had changed places at the Deluge (1700).

Thus a fruitful dialectic could be set up between data and interpretation, with limitations of vision being successively superseded, and better-directed observation opening horizons onto new problems. In this way the age's achievement was not a paradigmatic theory. It lay rather in a growing grasp of the *agenda* of any theory adequate to bear the load of the manifold functions it had to serve. The pressing *questions* – rather than the *answers* – for a theory of the Earth were becoming more clearly defined. I shall now examine in greater detail three examples of accumulating factual knowledge and expanding vistas.

Graduating from St John's College, Cambridge, in 1694, Abraham de la Pryme became curate of Broughton, near Brigg, in Lincolnshire. From then to his early death in 1704 his diary shows him as a diligent natural history observer (1869). He read the *Philosophical transactions*, was familiar with the theories of Hooke and Woodward, corresponded with Sloane, and used Lhwyd's *Lithophylacii* for identifying fossils (1869: 235f.; 1700: 682). He listened to folklore of fossils and landforms, but was sceptical (1869: 89, 106).

His attention focused on the problem of buried objects – fossils, skulls, ships' timbers found inland, coastal villages encroached on by sands (1869: 89, 91, 122, 137, 142, 148). This concern possibly sprang from his family's involvement with fen drainage, and certainly harmonized with his own antiquarian pursuits. His diary observations show a keen eye for evidence of migrations of land and sea (1869: 155).

The theoretical problem in natural history that puzzled him was the location of fossils. He had no doubt of their organic origin. His solution sprang from his own observations. The features which caught his eye were that fossils were frequently found in a damaged and broken condition; that fresh-water fossil shells were found embedded in different strata from marine remains; and that the strata containing them, consisting of silts, clays and sands, bore close resemblance to the materials in which shell creatures presently bred (1700: 677, 681f.). The former observations could not be accounted for by Woodward's theory. De la Pryme evidently thought Hooke's theory of continuous interchange of land and sea due to earthquakes and volcanic action was fanciful and lacked hard proof (1869: 106). Hence he advanced a theory in which at the Deluge much of the former sea-bed had been transformed into land and vice versa. This accounted for his three major observations. In modified form it became one of the most popular eighteenth-century integrations of fossils within Earth history (1700: 684f.).

John Ray's career presents the clearest example of a lifelong naturalist venturing a theory in his mature days. Ray had been successively severed from the pulpit, essentially by the Restoration religious settlement, and from fieldwork, by illness, family commitments, old age and poverty. Hence his theory was both substitute for, and completion of, his religious mission and his natural history. Whereas Abraham de la Pryme's theory flowed essentially from personal observations, Ray's life shows theory as the mature distillation of many kinds of experience, reading and reflection on the possibilities and limits of a scientific realization of the Earth. Raven having already established an excellent chronological description of Ray's thought, I shall merely draw threads together (1950: ch. xvi).

Ray's first known thoughts about the Earth are a sermon on the dissolution of the world, delivered in Cambridge in the 1650s (Raven, 1950: 442). It drew essentially upon theological learning. On his travels round England in the 1660s, he listed fossils and their locations, but chiefly with a collector's interest (1950: 421). On his subsequent Continental travels his more graphic experience of strata (at Bruges) and fossils (glossopetrae on Malta: 1738: 96–110; 251–7), provoked him to further thought, and stimulated the lengthy digressions, especially on fossils, in his *Observations topographical, moral and physiological* (1673). From the 1670s to his death, Ray corresponded with friends such as Lister, Tancred Robinson and Lhwyd, identifying fossils and particularly rehearsing, probing and revising the arguments used in the *Observations* for the organic origins of fossils. In 1692 he published his *Miscellaneous discourses*. This had for its core the above-mentioned sermon. Its theological learning remained much as in the 1650s, but its theme was now illustrated by extensive natural observations of changes in landforms. It included a digression on the Deluge, which debated fossils at greater length than before, and another on the Creation.[16] It discussed most of the important British and Continental writing in this area published over the previous thirty years, as well as many *Philosophical transactions* articles, and some from the *Acta eruditorum* (Raven, 1950: 442–3). It was recast as *Three physico-theological discourses* in the next year, when discussions of Creation, Deluge and Dissolution were now arranged in that order to form, for the first time in Ray's work, a comprehensive theory of the Earth. The two subsequent editions which Ray published before his death were further enlarged and revised with newly available evidence (Raven, 1950: 432–40).

Thus Ray's own observations, and those of his times, became woven into a theory in the 1690s. Yet though he formulated a total interpretation, Ray, the natural historian, remained deeply aware of the ambivalence of the evidence on such major issues as the origin of fossils. Furthermore, he believed that while Creation and Deluge were crucial hinges of Earth history, experience offered no unambiguous account of them – neither their causes, nor effects. Nor did observation provide a basis for a physical understanding of the Dissolution. Ray's *Discourses* are actually an *anti*-theory. While purporting to biographize the Earth, they demonstrate how such a project was impossible.

My last example of the interface of empirical and theoretical traditions is Robert Hooke. With de la Pryme and Ray, fieldwork was private, theory a public offering. In Hooke's case, however, form and audience were constant, but ambitions changed over time. Like his *Micrographia* (1665), his early 'Discourse of earthquakes', delivered to the Royal Society (1668, published 1705), began with a close empirical examination of figured stones (1705: 279–328) from which he inferred their organic origin (280–90). He argued on empirical, rational and Scriptural grounds that only successive operations of earthquakes could account for such fossils becoming embedded in the strata. Hooke's speculations on earthquakes were at this stage a bold, essentially *ad hoc* device introduced to justify his interpretation of fossils (1705: 290–328).

By the time of his lectures of the 1680s and 1690s, however, Hooke had incorporated the new empirical findings of the intervening twenty years – above all, reports of earthquakes and the discovery of giant fossil bones – within the ambition of a synoptic theory of the Earth (1705: 329–450; cf. Kubrin, 1968). He now saw the deep role of earthquakes within a structured, integrated Earth history, stretching from a pristine golden age up to the present, undergoing a continuing process of old age and decay (1705: 378f.). He investigated geo-physical causation for earthquakes and integrated the physics of the Earth's adopting its oblate spheroid form and of its changing centre of gravity (1705: 350f.; 372f.). He sought to legitimate his pattern of Earth history within the traditional wisdom of the Ancients (1705: 402f.). Davies has written 'Hooke's later geological writings display little of the perspicacity that is evident in his earlier work' (1964: 497), but this is to miss the point: his explanatory ambitions had changed.

### Theories, controversy and their audience

Theories of the Earth triggered off volleys of ripostes and counterblasts. Burnet's *Sacred theory of the earth* induced Erasmus Warren's *Geologia*, to which Burnet immediately replied in his *Answer to the late exceptions* (1690). Warren's further counter, his *Defence of his exceptions*, provoked his *Short consideration of Mr Erasmus Warren's defence of his exceptions* (1691). Opposition did not stop there. In 1693 John Beaumont added his own demolition of Burnet, *Considerations on...the Theory of the earth*, and Archibald Lovell donated his mite in 1696, in his *Summary of material heads*. To neither of these did Burnet deign to reply, but when John Keill entered the lists, with his powerful

*Examination of Dr Burnet's Theory of the earth* (1698), Burnet answered with his *Reflections upon the Theory of the earth* (1699). Woodward and Whiston elicited the same frenzy of theory and counter-theory.[17]

This controversy was no mere dispute within a narrow scientific community. It was markedly public scientific debate because its science operated within common-sense concepts. The debate drew in a wide range of 'outsiders', such as Herbert Croft, Bishop of Hereford, and Erasmus Warren, rector of Worlington in Suffolk, taunted by Burnet for his 'country philosophy' (1691: 16) – as well as those from other areas of science and scholarship. Further-more, the theory of the Earth became a talking point for the edu-cated. The age of William and Mary, and of Anne, was one of bitter polemics on the entire spectrum, from religion and politics to medical remedies (Holmes and Speck, 1967). In such a climate men readily seized pens to defend their theories, or to attack others', and the town eagerly bought tracts hot from the press to read and debate. In 1696 the Deluge stood as 'the Subject of most of the Philosophical Conversations of the *Virtuosi* about the Town' (Wotton, 1697: 66).

In this section I shall explore from two angles how controversy over theories of the Earth characterized and shaped the state of the science: relations amongst the theorists themselves, and be-tween them and their social audience. The problems of interpret-ing controversies are immense, and I can no more than scratch the surface. I shall examine two examples, the debate between Burnet and Warren, and that concerning Woodward's *Essay*, in the light of D. C. Allen's interpretation of the Renaissance humanist–scholarly debate on the Noachian Deluge. Allen estab-lished the high quality of historical and philological scholarship in this *querela*, but showed that in the course of two centuries the debate was neither resolved nor even focused upon a particular difficulty, because of the intractability of the written records and the impossibility of achieving agreement on a common method (1949). Would scientific debate, with an appeal to the objective tribunal of Nature, be any different?

Between 1690 and 1691 Erasmus Warren produced two attacks on Burnet's *Theory* and Burnet two defences. Warren's *Geologia* (1690) did not rise to any comprehensive assessment of the problems which Burnet had addressed, or any general review of Burnet's solutions. His was rather a string of *ad hoc* criticisms – many *ad hominem* – against particular parts of the theory, not all consistent with each other (1690: 23f.). The tone of Warren's attacks was polemical and forensic – Burnet called them 'wrangling and scolding' (1691: 1–2). Warren admitted 'my pen growing warm, quite out-run the bounds of my first Intentions' (1690: 'To the Reader'). The whole dispute was a characteristically logic-chopping, 'witty' altercation between scholars, with mutual censuring for their immoderate language and tone.

Though Burnet had adumbrated specific physical mechanisms, controversy largely bypassed these. Warren was chiefly concerned with refuting Burnet's interpretation of Scripture and classical authors, and Burnet of course replied on these issues. '. . .the *Theorist* has assaulted *Religion* and that in the very *foundations* of it', wrote Warren (1690: 'To the Reader'). The aspect of the *Theory* which most rattled Warren was its implication for man. For example, he was less at pains to refute on physical grounds Burnet's notions of original enclosed seas than to deny this was consistent with the dominion God had given over the animal kingdom (including fish).

In broad perspective, the controversy was fruitless. Warren had failed to grasp many of Burnet's original conceptions. Burnet accepted none of Warren's criticisms; nor did he modify his theory at all. Warren's second *Defence* and Burnet's reply danced the same steps with greater exasperation. Yet the dialectic of the controversy was highlighting certain real problems, though neither combatant might have recognized this. Warren castigated Burnet's physical extravagance over the Universal Deluge. But when Warren tried to explain it, denying himself a generous creation of water and Burnet's utter destruction of the Earth, he found that the waters failed to cover the mountains, and concluded that it had killed off mankind by starving, rather than drowning (Burnet, 1691: 13f.). Such an impasse generated a heightened awareness amongst their audience that the problem needed to be reconstituted. Likewise, Burnet and Warren clashed on whether mountains were beautiful, functional or mere ruins (Burnet, 1690: ch. x). This also sensitized opinion to an issue which was to become fundamental in the next century. The

Burnet–Warren controversy indicates that the chief passions sparked by theories were overtly religious, moral and humanistic, rather than physical. Men with incommensurable convictions were talking past each other, and using debating techniques which banished detailed and dispassionate technical examination of physical evidence.

John Arbuthnot's critique of Woodward's *Essay* is interestingly different. Arbuthnot and his friends were later to be scathing critics of Woodward as a man (Beattie, 1935; Nokes, 1975), but his *Examination* (1697) avoids *ad hominem* attacks, and presents itself as a constructive attempt to evaluate which aspects of Woodward's theory might be of lasting value to understanding the Earth. Thus, even though he established that Woodward plagiarized from Steno, Arbuthnot's point was not simply to condemn his intellectual pilfering, but rather to make a lengthy (and judicious) comparison between their ideas, showing where Woodward had improved on Steno, and the points at which Steno remained more plausible.

Arbuthnot began by summarizing Woodward briefly but accurately (1697: 1–10). Instead of wrangling over the details of the *Essay*, he launched into a methodological and philosophical critique (10–28). He showed that Woodward's theory appealed to events outside the present laws of Nature. But, except where miracles were specifically sanctioned by Scripture, he argued, philosophy should not go beyond known laws of Nature to explain the phenomena (8). Steno's theory was preferable in requiring fewer breaches in the laws of Nature than Woodward's. Furthermore, Woodward's theory seemed empirically dubious. The testimony of the rocks suggested that different strata were formed gradually over a period of time, not simultaneously, as Woodward insisted (24–5).

Arbuthnot's *Examination* shows that criticism which was perspicacious from the viewpoint of the advancement of scientific knowledge and methods was possible. It was, however, exceptional. The other writings in the Woodward controversy, especially the pseudonymous *Two essays* by L.P. (1695) and John Harris's *Remarks* (1697), were all fine words and invective, casting aspersions upon the character and scholarship of their targets rather than examining physical and methodological problems.

Whereas for Arbuthnot, Woodward's alleged plagiarism from Steno initiated constructive comparison, from the *Two essays* it drew personal attack. Harris sycophantically lambasted Woodward's detractors as 'invidious and morose', 'the Pest and Complaint of all Ages', 'Discouragers and impeders of Learning and Knowledge'. They had not 'offered one Objection of any weight', or 'invalidated so much as one Single Article of any of the numerous Propositions the Doctor hath advanc'd' (1697: Preface). The period of controversy was fruitful, though at one remove. Though not producing a victorious theory, it did expose the inadequacies of existing ones.

The theories aroused public interest, as is evident from the number of editions published, and also from the apparently large number of copies in some editions – Whiston claimed that 1500 copies of the first edition of his *New theory* were printed (1753: i, 38; V. A. Eyles, in Schneer, 1969). Those who did not read the tracts themselves were brought face to face with their authors as popularized in the *Spectator* (1911 ed.: i, 240; ii, 231), or as anatomized by the wits and writers of the day. The theory of the Earth became so well established in contemporary mythology that Swift saw to it that the ivory-towered Laputans in *Gulliver's travels* were discussing Whiston's Deluge theory (1906 ed.: 152f.). The audience at John Gay's *Three hours after marriage* encountered Woodward elevated into demonology as Dr Fossile, and found a play within the play on the Noachian Deluge (Beattie, 1935: 232). The public clamoured to see fossil shells and giant's bones in raree shows, and paid to hear Whiston lecture on the end of the world.

Theories of the Earth had been almost unknown in mid-century, but were popular science by its end. What does this mean? It shows an impressively rising general interest in grasping the natural world, particularly the human environment, through scientific investigation. This in part represents a displacing of Christian millennial expectation from political and eschatological dimensions onto Nature (Jacob and Lockwood, 1972). With very few exceptions, the reaction of the literate public to these theories was not one of horror. The Scottish Church apparently burnt Whiston's *New theory* – but that was mainly because of his Arian theology (de la Pryme, 1869: 159). There was no equivalent to

the belief rampant in the age of the French Revolution that Earth
science was set fair to undermine religion and society. For the late
seventeenth-century theorists were essentially working *within* the
common-sense assumptions of popular cosmology. The Boyle
Lectures show that the main bogeys were philosophical materialism, determinism and atheism, not theories of the Earth (Jacob,
1976). Fossil collectors were sometimes ridiculed as neglecting
human life for their collections, and theorists were satirized as
grandiose system-builders. But satire was directed more against
over-enthusiastic indulgence than against the science *eo ipso*, and
was festooned over the eminently parodiable personal characteristics of Woodward and Whiston, and the sitting target of the
wealthy, fashionable physician Sir Hans Sloane (Stimson, 1948:
130f.). Scientific satire in the period before the Glorious Revolution – e.g. in Shadwell and Butler – was almost silent on study of
the Earth, whereas Woodward and Whiston became two prime
targets after 1700.

Theories captured the imagination because they naturalized
and rendered scientific the central topics of human life, integrating basic issues of cosmology, theodicy, and the nature and place
of man. They encouraged pious cosmic meditation. At this stage –
contrast the early nineteenth century – the impact of theories of
the Earth upon the public was to open up edifying religious and
moral vistas. Public interest was not in minutiae. Theories did not
send *Spectator* readers scurrying out classifying fossils or charting
the strata.

Let us take stock of developments in the direction of Earth science
to the early eighteenth century, which marks something of a
watershed. No unanimity had been achieved on a theory – or even
on the particular issues within one: at most, a growing, if inex-
plicit, agreement on the agenda for solutions to the problems of
the Earth.

Active investigation had grown from negligible origins in the
mid-seventeenth century into lively scientific and public controversies early in the eighteenth. Every quantitative index shows
growth. Although no immediate, direct and special social *stimulus*
generated this increased preoccupation with the Earth, nevertheless society seems to have been becoming more *receptive* to natural

knowledge. Religious disputes were being taken before Nature's tribunal.

The intelligentsia was gradually but perceptibly redefining the criteria of wars of truth, away from scholarly techniques and the authority of books towards the norms of science and the objectivity of Nature, as is marked by the Ancients versus Moderns debate. Communities – albeit precarious and unstable – had arisen with shared interests in investigating the Earth. Natural history techniques, expanded into fieldwork on the mineral kingdom, were opening up new perspectives such as problems of strata. Field observation, especially in county natural histories, was beginning to make the *relations* between traditionally distinct fields of investigation study objects in their own right.

Theories of the Earth at the end of the century still bore strong family resemblance to the older cosmological and cosmogonical foci of natural philosophy and Biblical scholarship. But the subject was being dissolved and reintegrated. Burnet, Woodward, Whiston and Keill chose to discuss only the Earth itself rather than the whole system of Creation. The Earth's *future* was, perhaps, being hived off into entirely different study. Natural evidence was becoming far more central. Tenuous relationships had been forged between observation and theory. How the essentially local, agglomerative natural history tradition might relate in future to the analytical, theoretical aims of the geophysical sciences, or how description of a timeless Creation might dovetail with the historical explanatory ambitions of the theorists – these were open issues (Greene, 1959).

Furthermore, the theoretical structures which had been developed themselves posed paradoxical contradictions. New constraints – observational rigour, Newtonianism, natural theology – had been militating against acceptance of a naturally active Earth. Hence theorists – perforce and by choice – came to explain past Earth change by unactualistic, and even miraculous, catastrophic events or periods. Stringent methodology and the desire to disenchant and disinfect Nature were dissolving the major acts of Earth history into mysteries. This view of the Earth's destiny as involving 'revolution and stasis' emphasized *history*, but almost spirited that history beyond the bounds of possible inquiry. At this time there was not only no agreement, but also no obvious way forward, on many of the fundamental issues. A major high-level reorientation of philosophy of Nature would be necessary before such dilemmas could be reconceived.

But investigation of the Earth had developed its own new momentum, in the field and in the mind. Greater integration of particular studies – fossils, minerals, landforms – within an over-all view of the Earth had begun. The meaning of fossils was assuming historical dimensions. The mechanical philosophy and the very aspirations of theories were generating confidence that the Earth's form and processes were regular, orderly and capable of being understood within natural laws: both as a conclusion of scientific inquiry and as the very condition of that inquiry. Empirical study of fossils and topography was beginning to repay that confidence by revealing empirical conjunctions. Even in the absence of theoretical consensus, patterns of meticulous work-practice were being established which became normative for the future.

# 4

## A deepening base: *c.* 1710–1775

### *Disintegration and reintegration*

Investigation of the Earth in Britain in the first three-quarters of the eighteenth century is generally said to have been in decline and is frequently ignored.[1] This is part of a wider view that the gamut of the sciences – indeed, almost all intellectual life – was suffering a dark age in Britain at this time. In British natural philosophy, until the recent work of Schofield (1970), Heimann and McGuire (1971), and Thackray (1970a), the period has been stigmatized as a sterile interval between Newton and Dalton. Philip Ritterbush became exasperated at English investigations of life – at best, 'overtures' to biology (1964). D. E. Allen has traced a lull in natural history between about 1710 and 1745 (1976: 18f.). David Douglas believed the eighteenth century a fallow period in English antiquarian and historical scholarship (1939: ch. xiii). Stuart Piggott, studying the leading antiquary and archaeologist William Stukeley, explicitly confirmed this view (1950: Preface). Historians of the Enlightenment, such as Peter Gay, have been harsh on currents of intellectual life in England, beyond self-styled cosmopolitans such as Edward Gibbon (1967–70). A powerful historiography has seen eighteenth-century England as benightedly Christian, while Enlightened Europe was throwing off its theological shackles; as irrationalist, conservative, preoccupied more with arid theological wranglings and Classical literature than with modern humanities and the sciences; an England enshrined in the trinity of Dr Johnson, William Blackstone and Edmund Burke (e.g. Schilling, 1950; Stephen, 1876).

*Prima facie*, investigation of the Earth may seem simply to have suffered decline. Institutions such as the Oxford and the

Dublin Philosophical Societies died. Though the Royal Society outlived its crises and critics, it ceased to shape English Earth science as it had in the age of Oldenburg, Plot and Hooke (Weld, 1848: i, 483f.). The Ashmolean was denuded (Nichols, 1817–58: iv, 456). Balfour and Sibbald's collection in Edinburgh was dispersed (Scott, 1966: 47). Woodwardian professors in Cambridge kept the collection together but – against the terms of Woodward's benefaction – failed to lecture (Clark and Hughes, 1890: i, 182f.). No individual threw himself into the science with the single-mindedness of Woodward or Lhwyd. The little community surrounding Lhwyd, Ray and Lister did not acquire new blood, but faded away after their deaths. Richardson's sons dabbled in collecting as virtuosi, but without their father's dedication (R. Richardson, 1835: xxxii).

Although several new theories of the Earth were produced, their framework remained much as established by Burnet, Woodward and Whiston. Indeed, these authors' works continued to be reprinted well into the century, and they could still be referred to in 1785 by the Rev. James Douglas as 'the most intelligent of our natural philosophers' (1785: 53). Debate over theories lost its fierceness after about 1710 (V. A. Eyles, 1969).

Recently, however, historians of intellectual, cultural and social life in Hanoverian England have been recognizing how the conventional picture is a partial stereotype. To see the period as shallow, pettifogging and insulated from newer and more liberal currents being developed on the Continent is in part a relic of nineteenth-century Romantics' reactions against their own past. The view has been reinforced by conservative historians and literary critics who have hankered after a mythical, golden-age eighteenth-century England: stable, pre-democratic, pre-ideological. The popular image of mid-century remains the world of *Tom Jones* and James Boswell.

Historical revision, however, is in train. Certainly, many ancient institutions – such as the universities and Church – were in decline. But this may be an index of a modernizing society, outgrowing old forms. Economic historians have always stressed the age's commercial and industrial transformation, but only now is it being understood how these generated basic changes in cultural and intellectual life. Historians have recently been pointing to the blossoming of self-help and self-education (Hans, 1951), and the rise of a knowledgeable general public participating in modes of cultural and intellectual life from which they had previously been formally or *de facto* excluded (Plumb, 1972; 1973; Neuberg, 1971). Above all, historians have been discovering the upsurge of indigenous, middle-class, provincial culture: practical, rational, scientific in outlook, and associated with commercial and industrial enterprise (Thackray, 1974; Shapin, 1971; Musson and Robinson, 1969).

These movements had a profound impact upon Earth science in Britain, both in its social foundations, and in the generation of ambition and ideas. For to see investigation of the Earth suffering decline in eighteenth-century England is to take a partial, static and two-dimensional perspective. While certain traditions decayed, others evolved alongside. One man's decay is another's advance. That controversy about theories of the Earth became less virulent *may* betoken less commitment, but it may also mean that the jealousy which possessed so many late Stuart men of science, including Hooke, Newton, Woodward and Whiston, had yielded to the more civilized, co-operative relations of an age of stability (Plumb, 1967), scientifically more fruitful, and to the institutionalization of regular criticism. The dissolving of a collection *may* indicate simple neglect; but it may also show that with the advance of systematic classification, old, haphazard collections were now obsolete. Extinction does not challenge evolution but is a necessary part of it. The eighteenth century sees the forms of Earth science being reassembled to fit the new ecological niches of the age.

To find no peaks of achievement and to conclude thence that it was a period of decline is to be victim of false historical expectations – in particular 'great man' interpretations of history. Much investigation was conducted within traditions of natural history fieldwork which had been established in the seventeenth century. Such 'normal science' (as we may loosely call it) was accumulative. Similarly, theoretical projections of geo-history had established well integrated structures by the end of the previous century, which – however unsatisfactory – were unlikely to be transcended before major intellectual upheaval occurred. Most eighteenth-century theories rang the changes on existing structures. In any case, to find the decline of scientific activity in need of explanation is to assume that it is normal for science to grow unceasingly, for associations to expand continuously, for collections to be augmented. But European history up to the eighteenth century scarcely justifies such expectations. Till then the norm across the whole spectrum of intellectual, cultural and scientific pursuits had been for periods of efflorescence, associated with particular geniuses, patrons, or generations, followed by decay (cf. for museums, Murray, 1904). The Renaissance cyclical view of history reflected reality. Strikingly novel, however, is the virtually uninterrupted growth of all those activities since the eighteenth century. I shall explore some of the social causes for this major shift.

## Popularization

All the processes intimately involved in agrarian and industrial development in the eighteenth century – the emergence of a distinctively bourgeois, utilitarian, provincial culture, the growth of organized leisure facilities, the diffusion of knowledge – stimulated broadening social interest, and then a deeper institutional basis for the cultivation of science in Britain. *A fortiori*, such developments were crucial for investigation of the Earth, since its social purchase had previously been so precarious, and since – as I have noted before – it is an 'unrestricted' science, which requires many practitioners active over a variety of terrain ([Playfair], 1811: 207). The popularization of science served as a catalyst for the fusing of geology later in the century. It was to enable social groups who had previously scarcely existed, or who had been *de facto* excluded from the pursuit of science, to become integral to geology. No less importantly, it was to provide a larger participatory audience for geological writings and activities. The quantitative dimension of popularization – more books, more societies, etc. – included a subtler element, in the spread of interest to new areas of the country and new social groups, which generated its own adaptive feedback in the reorientation of ideas (cf. Mornet, 1911).

Traditional institutions were themselves not so moribund as is sometimes imagined. The Royal Society's *Philosophical transac-*

*tions* published a stream of important papers on the Earth sciences throughout the century. Emmanuel Mendes da Costa, the mid-century clerk, librarian and museum keeper to the Society, was especially energetic in orchestrating fossil studies through the country and promoting lecturing on the subject.

At least one Ashmolean keeper, William Huddesford, was active in the science. He reorganized the collection, and published a second, corrected and augmented, edition of Lhwyd's *Lithophylacii*. Though the Woodwardian professors in Cambridge did not lecture, two were of some eminence. Charles Mason extended the collection, displayed it to students and visitors, and made extensive scientific travels, collecting fossils and enumerating the strata. He kept in close contact with the leading naturalists and antiquarians of the day. His successor, John Michell, made fundamental contributions to the physical understanding of earthquakes and of strata, and, after he resigned his professorship to marry, undertook field traverses across England.[2]

Eighteenth-century science, however, grew most dramatically away from the metropolis and the ancient universities. A comprehensive series of societies was successively founded to cultivate agricultural, technological and scientific improvement as well as rational and polite knowledge. Directly and indirectly these societies encouraged knowledge of the Earth, for example by providing audiences for books and lectures, or by stimulating activities which required some expertise, such as collecting, agricultural improvement, local industries, mining, chemistry and landscape painting.

Gentlemen's Societies were among the first to be set up, from the 1710s onwards, centring first on market towns in Eastern England – Spalding, Peterborough, Boston, Lincoln (Moore, 1851; Winter, 1939; 1950). They stimulated reading, built up libraries, established collections of rarities and encouraged discussion of topics such as land-drainage and surveying. They were the precursors of the agricultural improvement societies and agricultural shows later in the century, whose importance lay in fixing landowners' interests firmly upon estate improvement and in cementing that alliance of landowner with surveyor or drainage expert which was so central to the development of practical geology (Bush, 1974).

For the relationship between industry and the rise of science in Hanoverian England has often been stressed. But we must not neglect the central role of land, agriculture and landowners in stimulating interest in the Earth. Landowners' income depended significantly upon exploitation of natural resources upon their estates – minerals and metals, coal, building stones, earths (Mingay, 1970: 163f.). The Board of Agriculture's county surveys at the end of the century produced the first mineral maps of many areas and Farey's exhaustive *General view of the agriculture and minerals of Derbyshire* (1811–15). William Smith and John Farey were to grow up within the landed economy as professional surveyors and improvers. Landowners were directly involved in the early years of the Royal Society of Edinburgh with the cultivation of geology (Shapin, 1974b).

Metropolitan societies also diversified. The many *ad hoc* scientific groups meeting in London, such as the Chapter Coffee House Club, contained men interested in the Earth – John Whitehurst, John Smeaton, James Keir, Richard Kirwan (Ellis, 1956; Musson and Robinson, 1969: 125f.). Slightly later, Sir Joseph Banks' *conversaziones* often turned upon mineralogical subjects (Cameron, 1956: 259). The refounded Society of Antiquaries contained many – including William Stukeley, Sir John Clerk, Roger and Samuel Gale – who made little distinction between natural rarities and human artefacts.

Of considerable importance was the founding of the Society of Arts in 1754, for its direct intention was to extend co-ordination between the improvement of science, technology, industry and agriculture, through offering premiums for specific improvements. One of the earliest prizes offered was for an essay on cobalt, and later the Society participated in promoting county mineralogical maps, the premium eventually being awarded to William Smith (Wood, 1913: 299f.). The logical extension of the Society of Arts was the founding of the Royal Institution in 1799, with its concern to bring together – though not *too* closely – fashionable society, professional men of science and artisans. In its early years, the Institution planned not merely to offer geological instruction, and to exhibit mineral collections, but also to house an enormous working mineralogical laboratory, as a national assay house (W. B. Jones, 1871: 142f.).

More important for geology was the totally novel origin of scientific societies in Midland and northern industrial centres in the second half of the century (Read, 1964). This movement, chiefly in the last two decades, will be discussed in chapter 6. But already by the 1770s the Lunar Society of Birmingham was flourishing, providing a model for later societies in establishing the importance of Earth knowledge to the needs of provincial culture, industrial and technological development, agriculture and transport improvements (Schofield, 1963b). Some later literary and philosophical societies simply provided an audience for geology, buying geological books for their libraries and entertaining geological lecturers.

But Lunar Society members actively promoted the science. John Whitehurst (1778) and Erasmus Darwin (1803) wrote full-scale theories of the Earth. James Keir produced a mineralogical and geological survey of Staffordshire (in Shaw, 1798–1801). Several members – including Keir, Wedgwood and Watt – undertook analysis of mineral products.[3] Wedgwood showed interest in fossils and strata while organizing the construction of the Trent–Mersey canal, and encouraged Whitehurst and Rudolf Erich Raspe to prospect for decorative spars and clays for pottery (Schofield, 1963b: 86f.). Boulton and Watt took a controlling interest in Cornish copper mining and set up an assay office there (Carswell, 1950; McKendrick, 1973). Members collaborated on such projects, sought mutual advice, and kept others informed of their activities. On an issue such as the basalt controversy, their individual interests and competencies fused into something like a distinct scientific interpretation (Schofield, 1963b: 174). Not least, Lunar Society members entertained many visiting geologists, such as James Hutton, J. A. de Luc, J. J Ferber, and Faujas de St Fond, and kept up extensive correspondence with other devotees at home and abroad. It goes without saying that, as before, individuals like Lewis Morris in Anglesey or William Brownrigg in Cumbria continued to study the Earth in their own private way (Morris, 1947). But the formation for the first time of provincial societies and *ad hoc* local working groups gave active stimulus to the science, and proved more productive in generating new ideas and publications.

The age of the commercialization of leisure (Plumb, 1973)

also developed numerous informal institutions for promoting and popularizing scientific knowledge. Privately owned, publicly accessible natural history and mineral collections and grottoes mushroomed. Famous, for example, were those of the Duchess of Portland and of Philip Rashleigh. Some also served as the basis for published catalogues, whether scientific, or lavishly illustrated picture books to entertain as well as to inform (Murray, 1904: ch. xii; Dance, 1962–8). Museums also began to flourish, again mainly privately owned but open to the public. The largest – Sir Ashton Lever's (called by Whitehurst 'incomparable' and 'perfect', 1778: 169), Coxe's, the Liverpool – were in London, but the provinces also supported several, such as Richard Greene's in Lichfield (Torrens, 1974b; 1974c). Most significant was the establishment of the British Museum, based on Sir Hans Sloane's bequest, and incorporating his collections and personal library. Despite frequent complaints about difficulty of access, the excessive speed of guided tours, and the lack of arrangement of the collections, the British Museum was the first publicly owned museum of its kind in Europe, open to any member of the public free of charge. It was a popular visiting place for the casually curious and for scientific workers alike. It housed impressive and growing collections of fossils and minerals, maintained by a succession of competent curators such as Matthew and Paul Maty and Daniel Solander, including the meritorious work of Charles Koenig early in the nineteenth century (W. C. Smith, 1969; E. Edwards, 1870). When the young Robert Jameson came to London in 1793 he significantly spent a great portion of his time visiting its rich profusion of museums and semi-public natural history collections, and the natural history and mineral dealers springing up at this time (Sweet, 1963).

Serving similar functions to societies and museums came the rise of libraries (Kaufman, 1969). These ranged from privately owned circulating libraries, run by booksellers, through proprietary libraries, book-clubs and subscription libraries set up by groups, to genuine community libraries. They all served to make an ample range of literature readily and cheaply available for the first time to a wide audience. Thousands of such libraries and book-clubs had sprung up by the end of the century. Some specifically served mining communities (Raistrick, 1938: 68). Despite contemporary

fears that mass novel-reading would swamp all other reading
tastes, even the circulating libraries chiefly stocked non-fiction,
amongst which were many scientific works, including books of
Earth science. Thus, borrowing figures for the Bristol Library
(Kaufman, 1969) show that Patrick Brydone's *A tour through
Sicily and Malta* (1773), with its discussion of the high antiquity
of the Earth, Goldsmith's *History of the earth and animated
nature* (1774), Buffon's *Natural history* (1749) and Hamilton's
*Observations on Mount Vesuvius* (1772) were all in strong
demand. Even a mere circulating library like Philpott's in Bristol
possessed volumes of Earth science, such as Buffon's *Natural
history*, Raspe's *On the German volcanoes*, and Maton's *Tour of
the western counties.*

Another expanding medium of popularization was scientific
lecturing to paying audiences (Gibbs, 1960). Viable first in London, it spread to the provinces. William Whiston gave famous –
or notorious – courses in London, as did Mendes da Costa in
mid-century (V. A. Eyles, 1969: 178). Johann Reinhold Forster
gave mineralogical lectures to the Warrington Academy. At first
the courses given by such itinerant lecturers as James Arden
(1772) in the provinces (especially the Midlands and the industrial North) were general introductions to natural philosophy or
chemistry, merely touching on the theory of the Earth or the
natural history of the mineral kingdom. By the early nineteenth-
century mineralogical and geological lecturers such as Robert
Bakewell and William Smith were emerging as specialists. Such
lecturing disseminated knowledge and stimulated interest, but
was also a significant source of notice, prestige and income for
surveyors, prospectors and mineral experts – a crucial feature in
the emergence of the free-lance professional consultant, who
could maintain himself financially by his geological skills.

All these forms of popularization presuppose a great upswing
in scientific publication. Particularly important was the explosion
of popular digests, compendia, encyclopaedias and text-books,
dealing partly or exclusively with scientific subjects, including
Earth science (Hughes, 1951; 1952; 1953). The explicit rationale
of so many eighteenth-century encyclopaedias was to diffuse
scientific knowledge. John Harris's *Lexicon technicum* (1704),
the first popular encyclopaedia, and James Hutton's first scientific

primer, carried extensive articles on the Earth by John Woodward himself. Successful later encyclopaedias, like Ephraim
Chambers' *Cyclopaedia* (1728 *et seq.*) and the *Encyclopaedia
Britannica* (1765–71 *et seq.*), carried lengthy discussions of rival
theories of the Earth, and individual articles on particular mineral
objects such as Granite, Sand, Talc, Mica. Subsequent editions
updated the knowledge, and organized the material into coherent
disquisitions on Cosmogony, Mineralogy and so forth.

Text-books likewise summarized knowledge about the Earth.
One of the most popular, Benjamin Martin's *Philosophical
grammar* (1735), contained an elaborate critical survey of theories
of the Earth, as well as lists of stones, fossils and strata. Martin
was one of many writers who were newly able to make a living out
of commercial scientific popularization. 'Sir' John Hill, another
scientific popularizer, published widely on minerals and fossils
within the whole range of natural history. His work contained
little original scientific research, his aim being to present existing
knowledge in a convenient fashion. In his *Fossils arranged according to their obvious characters* (1771) he used an external-charac-
ter analysis for identifying minerals, because it was most appropriate for those with little scientific training, and especially no
chemical skills (St Clair, 1966: 111; W. C. Smith, 1976). The
*Critical review* looked upon popularization's other face and sym-
pathized with Hill, who had been 'born a naturalist' but such
was the 'trifling spirit of the age' that he had been compelled to
become a hack, 'an universal scholar, a critic, polemic, casuist,
physician, metaphysician, politician, essayist, poet, novelist, and
astronomer, degenerating from, perhaps, the best botanist
Britain ever saw to one of the worst of her writers' (1759, viii:
272).

Emmanuel Mendes da Costa (1757a), Johann Reinhold Forster
(1768) and George Edwards (1776), among others, digested available knowledge on minerals and fossils into a convenient form.
They too made a living on the commercial margins of science.
These texts had some value in synthesizing the known, and bringing Continental writings before a British readership. The emergence of a readership for such works was beginning to enable men
without private means, a university education, fellowship or
benefice, to pursue a scientific career. The rise of the professional

expert living on his knowledge and skills is one important strand in the growth of geology (Pollard, 1965).

Translation is the foreign service of popularization. Cronstedt's and Bergman's mineralogies were translated (1772; 1783). Rudolf Erich Raspe, who supplemented his income by translating Continental geological works, proposed a regular translation agency (Carswell, 1950). He rationalized his own career as a scientific popularizer by arguing that science and industry could flourish only when philosophers and practical men made contact with each other through the medium of popular publications (1777: Preface). Raspe himself fused many currents of popular interest in the Earth in the latter part of the century – sentimentalism, the picturesque, the promotion of industry.

Reaching an even wider readership came the enormous growth of periodical literature. *The Spectator*, for example, gave reports of Burnet's *Theory of the earth* aimed at rational amusement and religious edification. Later in the century journals specially written for female audiences simplified geological theories (Meyer, 1955). From mid-century onwards the *Monthly review* (1749) and the *Critical review* (1756) gave serious regular accounts of new books on the Earth, under their policy of covering all new publications (Nangle, 1934). The reports were designed primarily as summaries, rather than as the kind of essay hung on a topic which the *Edinburgh review* pioneered early in the nineteenth century (Clive, 1957). Despite the *Critical review*'s conservatism, it treated scientific works to which it was hostile with amused indifference rather than with *odium theologicum* (e.g. 1758: 130). The *Monthly review* was a more earnest supporter of science and improvement. It also gave extensive notice of new Continental writings on the Earth and carried geological articles as well as reviews. The first English accounts of le Cat's theories were published in the *Monthly* (1750, iii), as were de Luc's attacks on Hutton. Brydone, de Luc and Whitehurst were occasional reviewers for the *Monthly*.

Three other kinds of popular writing blossomed, creating new perspectives on the Earth. From Ray and the early Boyle lecturers, through Paley to the Bridgewater Treatises, a very high proportion of Englishmen obtained their scientific vision of the natural environment through natural theology. In offering an illustrated guide-book, natural theology legitimated absorption in external Nature, turning the countryside into a temple of edification and instruction. But natural theology also educated the eye in substantive interpretative issues: the origin of the Earth, the function of mountains and rivers in the economy of Nature, the theodicy of fossils and extinction. With its concepts of adaptation, order, function and design, natural theology offered a book of the Earth which all who wished could read.

The second kind of popular literature reflects the surge of interest in all forms of topography – geographical works, local natural history, antiquarian studies, guide-books, maps, scenic prints, landscape painting, and above all, the gigantic literature of travel (Mitchell, 1900–1; Fussell, 1935; E. G. Cox, 1939–49). It was deeply rooted in contemporary society. Travelling for pleasure expanded greatly during the century, both within Britain, on the Continent and on voyages of exploration throughout the globe, facilitated by (and reciprocally creating) improvements in roads, transport, inns (Bayne-Powell, 1951; E. Moir, 1964). Tourism and holidays became institutionalized. Resorts sprang up in scenically spectacular mountain areas, especially near curative springs. In mid-century the Peak District was the major centre. Guided tours of Derbyshire caves were offered to musical accompaniment. Souvenir shops sold mineral items, from pieces of polished Blue John to systematic mineral and fossil collections (Ford, 1960). By the end of the century, the Lake District and Scotland were becoming fashionable (Nicholson, 1955). In addition, regional improvements, local pride and provincial economic promotions also sustained interest in local landscape.

At the very least, such literature stimulated and confirmed a trend towards the exploration of the countryside, which was to encourage amateur geologists. Furthermore, in the work of influential travel writers such as Bowles, Shaw, Pococke, Forster, Brydone and Thomas Pennant, topographical writing developed a high degree of interest and proficiency in discussion of landscapes and rock formations (e.g. Pococke, 1888: 101). Men like Pennant were able to make a living from travel literature, while also busying themselves on more specialized scientific studies. The popularization of Earth science made it easier for a professional community to emerge.

Lastly, the vogue for scientific poetry and landscape painting created another important medium for spreading – and thereby transforming – knowledge of the Earth (Nicolson, 1959; Parris, 1973). Scientific verse popularly combined natural religion, cosmic piety, description of landscape and detailed scientific fact, often in lengthy footnotes (Aubin, 1936). Early in the century, influenced by Newtonianism, poets usually treated the Earth in a total, cosmic context, focusing on cosmogonies. Later, with the fashion for Linnaean natural history and the growth of the aesthetics of the picturesque, attention turned to landscapes (W. P. Jones, 1937; 1966). A typical popular travel book early in the century, such as Defoe's *Tour* (1724), had little interest in natural landscape, finding non-cultivated land simply dreary waste. But by the time of, say, Richard Pococke in mid-century landscape was becoming interesting in its own right. In the age of Gilpin it became picturesque. The rise of this sublime and Romantic aesthetic appreciation of scenery is inseparable from that rich form of analysis of the Earth in terms of great antiquity, the majesty of slow and profound process, the investigation of subterranean depths, which is the geological way of seeing.[4]

I have surveyed that broadening social and institutional base for science developing in the eighteenth century which particularly gave impulse to examination of the Earth. This ensured for the future a wider readership for geological books or a larger audience for lectures. It helped to create the gentleman scientist–traveller and the amateur fieldworker, attached to informal local scientific networks or a member of actual societies. Not least, these developments helped to provide a base for the new kind of men who were to be the geologists of the future. Popularization did not in itself directly create a new science, 'geology'. Rather, it helped to create a taste for external Nature – both as an aesthetic experience and as rational exploration and local pride. Landscape had lost its terrors, and was becoming a kind of scientific playground, open to all. Furthermore, as I hope to show, the broadening, and changing, context for Earth science brought with it a certain shift in object, problems and emphasis.

## Earth science and the Enlightenment

Although the leading thinkers, writers and scientists in the Enlightenment were a fairly narrow group, the minds of the European intelligentsia at large – precisely those discussed in the previous section – were generally following the same direction. Their attitudes were becoming 'enlightened' (to use that word neutrally). The Enlightenment was a house which had many mansions, but also a certain unity.

The Enlightenment stood for suspicion and hostility towards religious cosmologies in which God acted voluntaristically, Nature was unintelligible, and man evil and helpless. Those which ascribed benevolence to God, order to Nature, knowledge and the potentiality of power to men were favoured. The heart of the strategy of the Enlightenment was the redefinition of values in the direction of a powerful *naturalism* (Willey, 1940; Haber, 1959). The extension of the concept and realm of Nature was not in itself anti-religious, but it did tend to hustle God further back, as merely the Author of a self-sufficient Nature. Nor was it anti-humanistic, since man was usually still conceived as in some crucial respects distinct from external Nature. But the thrust of the Enlightenment did have the effect of partitioning the natural more clearly from the divine and human, and endowing the natural with a pervasive reality. Enlightenment naturalism is in this respect the expression of a society and culture which had sufficient confidence to be able to insist upon a more highly ordered, rational view of the cosmos than traditional Christianity could provide, and sufficient control of its circumstances not to need to shun Nature in fear. But it was also a society in which Nature must still be thoroughly respected, because of the very magnitude and power that could produce the Lisbon earthquake, in the wake of which men still felt precarious and fragile.

Enlightened attitudes had a marked, if subtle, effect upon ways of seeing the Earth. They fired basic enthusiasm. For, as more of reality was redefined as natural, science thereby became the archetypal approach to studying it. But farther, Enlightenment ideas served to change notions of the scope and methods of Earth science, and also demanded reconceptualization of the Earth within its new perspectives on God, Nature, man and history.

How did the characteristic themes of the Enlightenment infil-
trate British thought about the Earth? It has been established that
the writings of Continental *philosophes*, from Bayle to the Baron
d'Holbach, were widely read in England, in original editions, in
translations and digested in reviews (Lough, 1970; Schilling,
1950; Kaufman, 1969). Some of these, such as Voltaire, Diderot,
d'Holbach, Nicholas Boulanger and Volney, explicitly discussed
the meaning of the Earth. Others made their impact through
general views on scientific method or the philosophy of Nature.
Furthermore, major Continental naturalists whose work was
coloured by the *philosophes* and by Cartesian rationalism found a
wide audience. Benoit de Maillet's *Telliamed* was translated
(1750). Long accounts of le Cat and Lazarro Moro appeared in
the reviews (le Cat, 1750; Zollman, 1746). And, above all,
Buffon's naturalism was influential, in the French original, in
translation, and digested into British works (e.g. Goldsmith, 1774).

But *native* British Deists, such as John Toland at the end of
the seventeenth century, had first championed free cosmic inquiry,
unrestricted by clergy or revealed truth, and the sufficiency of
Nature (Torrey, 1930). Throughout the next century this continued to generate a lively native tradition of natural philosophical
inquiry into cosmic and terrestrial issues. For his Cartesianism,
Thomas Burnet himself was viewed as what Ray dismissed as a
'mechanic theist', and as a free-thinker; Halley likewise.[5] Later
in the century the writings on the Earth by David Hume, James
Hutton, George Hoggart Toulmin and Erasmus Darwin, or in
poetic form by James Thomson and Mark Akenside, kept this
tradition very much alive (Dobrée, 1954; Sewell, 1961). Equally,
British philosophers, and especially Locke, had developed in
the late seventeenth century a psychologically-based empiricist
epistemology, extended through the next century as the basis for
a philosophy and method of science (Buchdahl, 1969: chs. iv–vi).
The characteristic preoccupations of Enlightenment natural philosophy and theology – the concern with order, intelligibility, regularity, process, Optimism – arose as part of an ongoing internal
dialectic within British thought as well as being stimulated from
across the Channel.

The Enlightenment sense of the profound reality of Nature
created awe in those attempting to unravel the Earth's nature,

history and purposes. As revealed religion's certainties about the Earth dissolved, and theories clashed blindly by night, the *limits* of scientific knowledge before the complexity of the Earth became the new strain. 'The present state in general, of most visible things may be discovered by a due and candid survey of them', the Rev. William Smith, describing the geology of the island of Nevis, confidently argued, 'But alas! To determine the means how they arrived at this state is in most cases too difficult a task for human understanding to go through' (1745: 248). Fieldwork helped to reveal a complexity to the Earth's operations and history which made the previous generation's neat geometrical theories appear fatuous. Sir William Hamilton's studies of Vesuvius showed not only the enormous number of lava flows, but also that the very nature and position of the volcano had changed. He did not know, and would not speculate upon, the causes of these changes or on volcanoes in general. He would merely offer 'a plain narrative' as he saw the facts (1768: 2). Mid-eighteenth-century literature on earthquakes was less ready to speculate upon theory, but more concerned to record empirical information, than equivalent seventeenth-century reports (Rousseau, 1968).

Discovery of the complications of stratification and of fossils led to favourable contrast of contemporary humility with the supposed overweaning rationalist ambition of the previous age. Oliver Goldsmith typically wrote:

It was the fault of the philosophers of the last age to be more inquisitive after the causes of things than after the things themselves. They seemed to think that a confession of ignorance cancelled their claims to wisdom; they therefore had a solution for every demand. But the present age has grown if not more inquisitive at least more modest and none are now ashamed of that ignorance which labour can neither remedy nor remove. (1774: 124)

The triumph of solid facts over flimsy theories was an Enlightenment trope. Richard Barton, discussing the origin of crystals, announced, 'Nothing shall be said with design to confirm or contradict, any former hypothesis of any other person', in order that the reader could have 'the pleasure of novelty without contentious dispute' (1751: 84).

Attachment to humble empirical knowledge encouraged attempts to reclassify scientific study of the Earth over and against disciplines built upon entirely different kinds of knowledge, or dismissed as bogus enterprises. The range of truth-authorities for Earth science was narrowed to the natural. Adherence to a *prisca sapientia* became discredited during the eighteenth century. Certain writers, notably George Hoggart Toulmin and James Hutton, discarded written human testimony as valueless for Earth history.[6] 'It is not in human record but in natural history, that we are to look for the means of ascertaining what has already been' (Hutton, 1785). Naturalistic investigation of the Earth became increasingly separated from scholarly chronological studies.

Furthermore, demarcations were driven between empirical investigation and other forms of inquiry – general cosmology and cosmogony, natural philosophy and natural theology, and human history, whose relevance was becoming questioned (Raspe, 1970: 2–3). Study of the Earth became divorced from the larger questions of celestial mechanics. Raspe, Hutton and Toulmin distinguished the science of the Earth from nebulous questions of cosmogony, and ignored the issue of the Earth's future. For Whitehurst, as, later, for Hutton, Earth science was a study of 'CAUSES TRULY EXISTENT' (1778: ii). Beyond the present laws of Nature, science ceased. Growing acceptance of the Earth's high antiquity ended the absolute indissolubility of Earth and human history, the latter tending to become merely the final stage of the development of the globe.

The Enlightenment drew boundaries between scientific investigation of the Earth and revealed religion. The Catholic–Cartesian double-truth strategy, that the Bible and science offered separate but equally valid forms of truth, found some, but not much, favour in Britain. Men such as Hutton and Toulmin were to dismiss the Bible as misleading and irrelevant to Earth history. More commonly, British Enlightenment thought either asserted that the Bible was not totally inspired, or that it was not intended to teach physical truth, or that Genesis must be liberally and rationally interpreted. 'I never yet heard it suggested before that these sacred Books were canonized as Treatises of universal History and Chronology' wrote Richard Clayton, Bishop of Clogher (1752: 45).[7]

One concrete accommodation of the Bible to fit the needs of reason and science was the increasingly popular view that the Noachian Deluge could have been only partial. A limited Deluge

could serve God's purposes equally well, and could even demonstrate a more humane and enlightened God (Clayton, 1752). Only
the pride and ignorance of the Jews made them believe it was
universal. A partial Deluge avoided the grave rational and scientific problems posed by a universal flood and could be explained
naturally without a miracle. Most mid-century Enlightened
Christians believed the order of Nature was a better argument for
God than breaches of it. Though British explanations of Earth
history remained religiously oriented, the religion became more
naturalized. Thomas Burnet's millennial, new spiritual kingdom
became William Worthington's physical improvement of the
earth (1773), and Erasmus Darwin's evolutionary progress (Tuveson, 1949).

Enlightenment methodology and epistemology, then, cemented
Earth science more firmly within empirical inquiry, and demarcated it from other sciences and forms of discourse. Put negatively,
this meant that Enlightenment patterns of thought exerted constraints to allow only a narrowed-down body of empirical observations to count as the true science of the Earth. Earth science, like
others, had got epistemological jitters. Put positively, it meant
that that body of observations and laws could be taken confidently as a *true* science of the Earth. For, present in the thought
of de Maillet, le Cat, Whitehurst, Goldsmith, Hutton and
Toulmin was the perception of the independent, and assured,
continued existence of the Earth with its own laws, free from
theological interference and destruction: here to stay.

This of course relates at an ontological level to the influence of
Enlightenment notions of Nature in reconstituting understanding
of the Earth. For fundamental to Enlightenment thought was a
growing belief in the goodness, perfection, self-sufficiency and
activity of the natural environment. This view was not necessarily
anti-Christian and certainly not anti-religious. It had many roots
and ramifications, but perhaps is best seen as a development from
those seventeenth-century rational Christians and natural
theologians who saw God's wisdom, goodness and power guaranteed in the order, regularity and lawfulness of Nature. They had
denied that Nature was decrepit, decaying, ugly or malfunctioning; but Protestantism and Newtonianism had combined to
emphasize Nature's passivity. But now an active, progressive

Earth capable of producing new forms, repairing its own losses, and even of supporting new varieties of creatures became a more admirable testament to God's – or Nature's – wisdom and foresight, and a more permanent habitat for man (Crane, 1934).

Eighteenth-century natural philosophy required active powers to solve the problems of force and matter. The life sciences needed similar conceptions to transcend Cartesian dualism and to grasp the variety, diversity and the fecundity of Nature (Figlio, 1975; Glass *et al.*, 1959). Enlightenment thinkers, then, began to view the Earth increasingly as a theatre of active processes. They began to accept that natural forces, working gradually over time, could progressively produce new continents, strata, mountains and rivers; and that opposed natural forces could be continuously destroying those very features. They could envisage all such forces acting within an integrated progressive economy of nature which, while permitting local change, creation and destruction, held all things in a stable or evolving equilibrium. Buffon's was the classic naturalist's exposition of this view, but it is found canvassed in Hume (1779) and, in a more eclectic form, in a great deal of eighteenth-century British 'stoic' cosmic poetry (W. P. Jones, 1966).

The static view of Creation was being challenged by a dynamic image of self-re-forming and self-sustaining nature (Greene, 1959). The Great Chain of Being was temporalized (Lovejoy, 1936: ch. ix). Alongside timeless, descriptive natural history associated with classification, a dynamic conception of natural history developed, the French *histoire naturelle*, the Scots 'conjectural history', in which the very word 'history' began to shed its prime seventeenth-century meaning of 'description' and to assume its more modern meaning implying the pattern of events over time (Greene, 1959).

The speculative theories of the Earth of the French Enlightenment all embodied these features. De Maillet, le Cat, Buffon, Boulanger and d'Holbach each developed Earth histories in which the natural forces of the terrestrial economy – tides, rivers, weather, denudation, sedimentation – continuously destroyed and rebuilt the face of the globe. Lengthy processes of consolidation of new strata on the sea-bed, and the gradual entombment of organic remains within them, were essential to these theories. Nothing in Nature was pristine. All was the child of time and process, i.e. Nature (Haber, in Glass *et al.*, 1959). In the next chapter I shall examine specific cases of how such ideas were – partially – incorporated in British formulations about strata and fossils.

Integral to the Enlightenment view of Nature's activity and sufficiency was a revolution in ideas about time. As the specifically Christian God receded, and the men of the Enlightenment laid stress on the reality of Nature, a high antiquity endowed the Earth with solidity, sublimity and continuity. A long – or indefinite – time-scale was the logical correlate to the previous century's acceptance of infinite – or indefinite – space. More liberal attitudes to the Bible could permit the view that Scripture had merely set a date to *man's* existence. Such a view had been Halley's and was mooted by Richard Watson, Bishop of Llandaff (1806: 153). To accept that natural processes were responsible for great revolutions of the Earth required a lengthy time-scale.

The revolution in time which fused the naturalistic and the romantic was twofold. In part it was forging high numerical estimates for the Earth's age. Buffon was a pioneer in this with his extrapolations from experiments with cooling iron balls (Haber, 1959: 115f.). But time also took on a new conceptual role, as a force in the economy of Nature. Once again Buffon was to the fore in creating such a view, adopted in Britain by Hutton, and by Playfair, with his analogy, from calculus, of time as the integrating power of Nature.[8]

Knowledge of the Earth, then, became more widely broadcast; and what was disseminated was an increasingly naturalistic perspective on the Earth's crust and upon landforms, which paralleled the rise of landscape painting and an appreciation of external Nature's beauties. An Enlightened commentator like Horace Walpole found notions of the Deluge amusing, and Josiah Wedgwood complained when John Whitehurst supposedly dragged Moses unnecessarily into his theory of the Earth (Schofield, 1963b: 177). Indeed the whole thrust of Alexander Catcott's *Treatise on the Deluge* (1761) was expressly to refute the naturalistic interpretations of the history and meaning of the Earth which he saw rife. But the classic embodiment of the popularization of eclectic Enlightenment views of the Earth is Oliver Goldsmith's *History of the earth and animated nature* (1774, and many editions; Pitman, 1924). Probably the most popular work of natural history in Enlightenment Britain, it fused both the specialized research of the *Philosophical transactions* and the Enlightenment philosophy of Buffon within a series of graphic views of, and hymns to, Nature.

Programmatically, Goldsmith affirmed that the vulgar either had no curiosity about the Earth, or attributed all phenomena immediately to God's goodness or wrath (1774, i: 1–3, 17–18, 87f.). Only recently had philosophical curiosity been kindled. But how should the Earth be approached? Goldsmith rejected previous speculative cosmogonies as over-ambitious, believing that rationalist or providentialist attempts to understand the Earth's functioning directly through final causes were ill-conceived (ch. iv, 18, 130f.). Nature should be understood not in relation to specific intentions of God or narrow needs of man but as a uniform, regular, overall economy of causes and effects. He believed that many problems, including strata and many aspects of fossils, were beyond current knowledge (ch. v). But he did stress the active powers of Nature in the denudation of mountains (140f.), and their reconstruction through volcanoes and earthquake activity, which gave Nature local mutability within overall stability (145, ch. xvii).

Goldsmith's work is a series of discrete views of Nature, rather than a connected work of science. It is not yet 'geology'. But it conceptualized the Earth as a stable, self-regulating natural economy, in a way which was foreign to late seventeenth-century thought but which was to be so very influential in the formation of 'geology'.

# 5

## Continuity and change

### *The momentum of normal science*

The forms of the natural history of the Earth such as were pioneered by Plot, Ray, Lister, Lhwyd, Woodward took root and proved extremely tenacious. They provided usable patterns and techniques with which investigators in Hanoverian England found they could fruitfully work. This does not gainsay the arguments just set out for a broadening social base, and shifting approaches to the Earth. For interest in the Earth was always multi-faceted, producing a range of disciplines and strategies. Investigation at traditional levels continued, while new forms were being forged, and such traditions were reinforced through continuing currency of earlier books, collections, catalogues and classifications, and not least by sustained circulation of *Philosophical transactions* articles, especially in popular abridged form (Lowthorp, 1705; Hutton *et al.*, 1809).

In particular, the two modes of natural history outlined earlier, the one concerned with topography, the other with cabinet identification and classification of specimens, had produced a momentum which encouraged routine expansion within the tradition. Local natural histories grew in number and popularity. Particularly detailed works were Edward Owen's *Observations. . . about Bristol* (1754; cf. S. Smith, 1946), James Pilkington's natural history of Derbyshire (1789), John Walker's 1760s surveys of the Highlands and Islands (not published till after his death (1808) and Lewis Morris's extensive (unpublished) investigations of the mineral and metal areas of Wales. As historians of a single area, usually a county, many worked within the same physical and conceptual limits as had Plot, or Leigh half a century earlier. Minerals, fossils, strata, hills and rivers still tended to be catalogued independently of each other, with no strong integrating conception of an organic structural geography.[1]

Nevertheless the mineral history of many parts of Britain previously *terra incognita*, including Wales, Ireland and Scotland, was now being recorded and published. Important for example were Sir John Sinclair's statistical accounts of Scotland, which collated local data about minerals, earths and topography (Mitchison, 1962). Likewise the Board of Agriculture's county surveys each included an account of local strata, soils and mineral resources. Though some were perfunctory, others, such as Joseph Plymley's *Shropshire* (1803) and above all John Farey's *Derbyshire* (1811–15), were of lasting importance to geology (Sir E. Clarke, 1898; Bush, 1974).

Furthermore, county natural history at its best displayed sensitive discrimination, precisely because of the author's intimate knowledge of his locality. For example, the Rev. William Borlase in his *Natural history of Cornwall* (1758) took a generally conventional view of the Creation and Deluge, with his natural theology of overall stability; but his familiarity with the devastations of tides and weather on the Cornish coast-line, and his close knowledge of old maps and accounts of the peninsula and the Scilly Isles, led him, almost despite himself, to a powerful awareness of denudation (1753; 1756). Similarly his expertise in Cornish mining forced him to recognize that the extraordinary complexity of intersections of veins, lodes and faults must mean a succession *in time* of their causes. Borlase conventionally accepted that the Deluge had produced the greatest strata disruptions. But his familiarity with the evidence required him to postulate subsequent catastrophic events as well.

Local natural history was dramatically augmented in this period by travellers' accounts. I shall argue later that the travellers' experiences were such as to bring new perspectives to understanding landscape structure. Here, I shall merely note the quality of their observations within familiar traditions of investigation. For example, Richard Pococke's account of the Giant's Causeway (1748) slotted into the series of articles on that problem published in the *Philosophical transactions* – as he actually acknowledged. Not only had he personally visited the Causeway – unlike certain

of the earlier writers – but also measured it, and sought to establish its relations to the adjoining strata. Five years later, after re-
visiting the Causeway, he published a map and a drawing, and
drew attention to its similarities with other basalt outcrops in the
area (1748; 1753). Work of this kind graduated into self-conscious
'volcanist' interest from the 1760s. It is hardly surprising that,
unlike earlier writers on the Causeway, Pococke had extensively
surveyed the landforms of Scotland, the Alps and other parts of
Europe (1744; 1891; 1888). Another traveller who produced
many books of travels was Thomas Pennant, noteworthy for his
combining of picturesque scene-painting with considerable precision of mineral description (cf. Matheson, 1954; W. C. Smith,
1911–13).

Naturalists were also actively consolidating these traditions of
natural history identification and analysis developed by the previous generation. A steady trickle of mineral classifications appeared
– those of Mendes da Costa, George Edwards, 'Sir' John Hill,
Johann Reinhold Forster – as also did works describing fossil
collections, or the fossils of a particular area. In this genre,
Gustavus Brander's *Fossilia Hantoniensia* (1766) was important
as the first British fossil treatise to employ the new Linnaean classification. John Walcott enumerated the fossils of the Bath area
(1779). And the tradition of Lister, Ray and Lhwyd was continued in a lengthy series of *Philosophical transactions* articles by
such competent naturalists and microscopists as D. E. Baker,
Gustavus Brander, Emmanuel Mendes da Costa and Joshua Platt
on difficult fossil problems, such as the belemnite or the reconstruction of giant vertebrate bones.[2]

For a short time, da Costa seemed poised to use his position as
clerk, librarian and museum keeper to the Royal Society to publish a mountain of useful information in Woodward fashion upon
the location of particular British fossils. He established an unrivalled correspondence, sending out questionnaires and requests
for specimens. Some of the information gleaned was eventually
used in his publications. His endeavours, however, were undermined by his expulsion from the Society, though he later published and lectured on fossils and minerals. Yet his scientific
ambitions were essentially those of the previous century: the
accumulatory strategy of natural history, the concern with speci-

mens. There is no intellectual revolution between Lhwyd and da Costa.[3]

Most eighteenth-century mining manuals also kept to existing traditions, rather than rejecting them. Indeed, this was their declared intention – the handing down of personal craft experience from generation to generation. In his *Miner's dictionary* (1747) William Hooson proudly boasted, *contra* the German technocrat Linden, that his was the true ancient mining lore, now written down for the first time just before it died out (Porter, 1973b; Rhodes, 1968). For deep social and professional reasons, there was no development of a science-trained elite corps of mining experts in mid-century, who alone could have transformed mining lore into a geological technology. The old lists of 'symptoms' were still recommended for discovering minerals and metals. These were augmented by extrapolations from a hazily inferred order of strata, but as strata had not yet actually been recorded over wide expanses of Britain, such knowledge was crude and often misleading, as those who bored for coal in Dorset, Norfolk, and Buckinghamshire found to their cost. Confidence in 'symptoms' – that there was an exact correlation between surface features (such as bad air, discoloration of streams, the outcropping of particular rocks) and mineral deposits – betrays a lack of real structural knowledge of the relations of deep-seated deposits to relief (cf. [Sharpe], 1769; Torrens, 1974a).

A growing number of miners' dictionaries were published, and these helped to standardize nomenclature and explain terms. More sections taken from mine shafts were collated and printed, as by Sharpe and Joshua Dixon, and could form the basis for later extrapolations (Dixon, 1801: 107; 113–25). Furthermore, though a book such as William Pryce's *Mineralogia Cornubiensis* (1778) appears a mere compendium of empirical materials when compared with Continental works such as Delius's *Einleitung zu der Bergankunst nach ihrer Theorie und Ausübung* (1773), it nevertheless provided a comprehensive digest of Cornish mining practice. Pryce described the mineral structure of the Cornwall metal mining areas, and discussed relationships between the various kinds of ore-bearing lodes, tabulating his knowledge in diagrams, sections and glossaries. The work is of an encyclopaedic assurance unknown in the previous century.

Broader interpretations of Earth history could still be accommodated within the long-established frameworks of the theories of Burnet, Woodward and Whiston. Indeed one consequence of the increasing diffusion of knowledge was perhaps to lengthen the life expectancy of the older theories. The familiar problems of the denudation dilemma, the universality of the Deluge, the local place of fossils, still largely dominated theoretical inquiry.

The Bristol clergyman Alexander Catcott (1725–79), for example, undertook two decades of dedicated fieldwork, working within the explanatory frameworks established by his late seventeenth-century predecessors.[4] His connection with those theories was a close, personal one. For Catcott took his ideas from his father, Alexander Stopford Catcott (1692–1749), who was a friend of the John Hutchinson who himself had been the paid fieldworker of John Woodward. Alexander Catcott strongly defended the main tenets of Hutchinsonianism as set out in Hutchinson's *Moses's Principia* (1724–7), that the system of the Earth was essentially a Divine, stable machine, and that all major landforms were to be attributed either to the original miracle of Creation, or to the universal Noachian Deluge, which had reduced the entire Earth to its atoms, and subsequently re-created it (A. J. Kuhn, 1961; Neve and Porter, 1977).

Catcott's extensive fossil collecting experience as an Oxford undergraduate confirmed his conviction in two cardinal aspects of the theory: that fossils were indeed organic remains, and that the strata were to be found superincumbent in regular horizontal order. Throughout his travels his basic expectation was to confirm his theodicy of 'order', i.e. the relatively horizontal position of the strata within a gently undulating countryside. For Catcott, such landscape demonstrated that, since the Deluge, no chance conjunction of natural forces had subsequently disturbed the Earth to any significant degree. More problematic were the instances of strata at considerable angles to the horizon, distorted, folded and faulted, or jagged landforms, such as the Avon Gorge, erratic boulders, pot-holes and caverns. He termed these features 'disorder', and attributed them to the Deluge, since no regular processes of Nature could have been sufficient cause. He believed they followed from the destruction of all strata at the Deluge, and were the effects of torrents rushing off the land. In most respects this Woodwardian and Hutchinsonian realization of the Earth offered Catcott a highly fruitful framework within which to pursue perceptive field studies. For, just as in the case of Mason, Michell and Whitehurst, Catcott could begin to break down the late seventeenth-century polarization of theory and fieldwork, and integrate the two within a comprehensive science of the Earth. However, as will appear, his position as a fieldworking traveller, and his empirical discoveries, created tensions in his adherence to traditional theory.

William Stukeley, the Cambridge-educated naturalist and antiquary, is another example of continuity of approach from the late seventeenth century (Piggott, 1950; Stukeley, 1882–7). Stukeley had two main interests in the Earth. The first was a passion for all objects found within it, be they human artefacts or natural remains (and often the nub was to distinguish them). His diaries and letters reveal an almost haphazard, virtuoso concern to accumulate such natural objects (Weld, 1848: 526–7). His extensive travels shew him exposing Roman remains and digging out fossils without feeling any essential distinctions between the two activities (Stukeley, 1724). The second interest was his desire to construct a sacred history integrating the Earth and man (Stukeley, 1882–7: 117f., 126f.). In man's past, the link between Biblical history and British history was the Druids (Piggott, 1968: ch. iv). In Earth history, the pivot was the Deluge, which (as for Catcott) explained most enigmatic phenomena. But then the Druids and the Deluge were linked, for the Druids had brought civilization to Britain after the Deluge, which their religious rituals memorialized.

Naturalists such as Stukeley and Catcott could be such staunch defenders of traditional shapes of Earth history precisely because they were self-appointed opponents of Enlightenment trends in religious and scientific thought (though in significant ways they were themselves beneficiaries of the Enlightenment). Stukeley resisted all Enlightened attempts to minimize the Deluge's role in Earth history (Stukeley, 1750: 642). Catcott's Hutchinsonianism was explicitly aimed against both liberal Protestantism and Enlightenment naturalism, and his theories of a universal and totally destructive Deluge were directly opposed to the rational Christianity of such as Bishop Clayton and the Enlightenment view of an active Earth as developed by le Cat and Buffon.

Thus within particular areas of Earth science, investigation in the eighteenth century continued to follow the Stuart pattern – for example in the extension of collections and the steady development of chemical and mineral analysis (W. C. Smith, 1911–13; Matheson, 1954). 'Normal science' of this kind within essentially descriptive natural history sciences was not aiming for intellectual revolutions; merely to pencil in the blank areas on the map of Britain, and to fill the data-bank with useful facts. Such continuity – particularly when it involved ever-widening circles of naturalists – does not represent a failure of nerve or imagination, but rather the successful establishment of routine scientific practices, which are the bread-and-butter of workaday natural history (Allen, 1976). The work which the geniuses of a previous age – Ray, Lhwyd, Lister, Plot – had had to pioneer could in the future be left to the average competent naturalist. The base was deepening.

### New trends

There were also new trends in ways of interpreting the Earth. To some extent these developed out of problems which were internally generated. To some extent they represented the viewpoints of sectors of society newly involved in natural history, or evolved out of other new developments such as more ambitious travel. To a large degree they were also coloured by new profiles of Nature encouraged by Enlightenment changes in attitude. All these pressures helped to guide inquiry in the direction of what eventually became 'geology'.

The most significant trend was growing conceptual and practical focus upon strata. There had been, of course, certain late seventeenth-century stimuli: for example, Woodward's and Hooke's theories, and the growth of mines observations. But the most important new impulse was the vast increase of scientific travelling. The problems posed by the succession of outcrops, the parallelism of the strata, breaks and faults, unconformities and so forth, only appeared as problems and became soluble once the investigator's gaze became fixed on the relations of continuities and discontinuities over extensive tracts of land. Even those seventeenth-century observers who had travelled widely had tended to focus their attention on quarries and other spot exposures, without any programme for studying and conceptualizing the intervening terrain as a three-dimensional *bloc*. Travelling across Britain with the specific aim of enumerating the strata did not become common until the eighteenth century in the work of such as Alexander Catcott, Charles Mason, John Michell, John Smeaton and Henry Cavendish.

Empirical study of the strata could be communicated in many forms. Perhaps the most basic was a perpendicular verbal or pictorial section, reproducing the exposure of a well or a mineshaft (Dixon, 1801; Challinor, 1954; Rudwick, 1976). More sophisticated, because projecting information across an *area*, was the kind of panoramic section pioneered by John Strachey (1719; 1725; Fuller, 1969). Strachey's section and his comment demonstrate that he recognized the discontinuity between the dip of the strata of the Somerset coal country and the Mendip lead mining measures. His sections show him describing these strata meeting unconformably – though it is not clear whether he saw any theoretical significance in this phenomenon (Tomkeieff, 1962).

The growing belief that strata were continuous over considerable stretches of terrain gave confidence to attempts to trace them. As a follower of Woodward, Benjamin Holloway had had his mind sharpened to the significance of the strata. He studied the strata of Bedfordshire with care, and, after comparing his findings with published strata data, concluded that the same beds appeared to run in constant order from the environs of Oxford across to Newmarket Heath, producing topographical continuity (1723).[5]

Charles Mason, Woodwardian professor in Cambridge, travelled extensively, particularly through the Midlands and Peak Districts, and he recorded outcroppings in his travel diaries. Like many horseback travellers, he learnt to infer the main rock structure from building stone and from the rubble used to make and mend roads. Alexander Catcott rode round the Midlands, the Peak District and through much of Wales between the 1740s and the 1770s, and listed the strata and made occasional sketch maps and sections of the stratification of hills, exposed cliffs and quarries. Like other travellers, he noted the conjunctions of particular rocks with distinctive fossils, surface vegetation and useful minerals, and made numerous *ad hoc* comparisons between the stratification of different areas. Particularly penetrating was his retracing of pebbles, gravels and other 'diluvium' to their original exposures (Neve and Porter, 1977).

Somewhat later, the scientific diaries of Henry Cavendish register numerous excursions taken with Charles Blagden, John Michell and John Smeaton, where recording strata was one of the prime interests (Harvey, 1971). Naturalists such as Catcott, Michell and Cavendish were recording details of strata more consistently and more accurately than their seventeenth-century predecessors (Geikie, 1918: 47f.). Yet, beyond recording, little use was generally made of such strata observations. Mason seems never to have worked up his observations into any interpretative form. Catcott seems to have been satisfied with proving the basic fact that the strata lay *in order*. So although scientific travelling made strata observations that much more important, conceptual interpretative frameworks for these observations lagged behind.

Some travellers, however, began to grasp how knowledge of the structure, continuities and discontinuities of the strata over broad areas of terrain might be the means to understanding the Earth's crust. Thomas Pennant, summing up his *Tour of Wales* (1772), attempted a generalized stratigraphical survey of the North Wales mineral area: 'I shall now', he wrote, 'take a kind of bird's eye of the whole, which I have surrounded in the course of my tour', seeking to divide up the area into 'limestone' and 'chert' districts and discussing the minerals – lead, calamine, zinc – to be found within each (1810: 64–74).

The output of John Michell, however, is the nearest approach at this time to an Earth science articulated around enumeration and interpretation of the strata. He was able to go beyond *strata*, and set out, perhaps for the first time, a clear theoretical definition of *stratification*:

The earth is not composed of heaps of matter casually thrown together, but of regular and uniform strata. These strata, though they frequently do not exceed a few feet, or perhaps a few inches, in thickness, yet often extend in length and breadth for many miles, and this without varying their thickness considerably. The same stratum also preserves a uniform character throughout, though the strata immediately next to each other are very often totally different. (Geikie, 1918: 41)

Furthermore, Michell could also conceptualize stratification into simplified, abstract thought-experiments:

Let a number of leaves of paper of several different sorts or colours, be pasted upon one another, then bending them up together into a ridge in the middle, conceive them to be reduced again to a level surface by a plane so passing through them as to cut off all the part that had been raised; let the middle now be again raised a little, and this will be a good general representation of most, if not all mountainous countries together with the parts adjacent, throughout the whole world. (Geikie, 1918: 42–3)

His was a *working* grasp, in that he could perceive the meaning of stratification for other problems of the Earth's crust – e.g. in his theory of the longitudinal propagation of earthquakes (Michell, 1760). And his grasp was many-sided, serving various purposes. He constructed lists of the general horizontal order of English strata (Farey, 1810). But, equally, he was able to trace, with considerable accuracy, outcroppings of the strata over wide areas and thence to generalize about the lithological structure of England.[6]

The popularity of topography and of scientific travelling during the eighteenth century made practical exploration of continuous tracts of relief possible, and attractive, and strata became the key to such study. Furthermore, strata study was given a conceptual boost as the object of Earth science was refined down from the globe to the Earth's crust, and Enlightenment naturalism bolstered confidence in the order, stability, grandeur and historical importance of the strata. Various kinds of inquiries into strata were being pursued. Knowledge was being accumulated of strata outcrops across the British Isles, and of the general structural features of strata (strike, dip, faulting, unconformity, outcropping). Elementary generalizations were being more frequently ventured – that certain strata regularly underlay others, some were fossiliferous or metalliferous; some associated with upland or lowland areas. Furthermore, in a profoundly important step for Earth science, investigators began to reconceptualize the strata. Few seventeenth-century thinkers had believed strata in themselves could reveal much of Earth history, assuming that strata were generally in no special order and had been created more or less contemporaneously at the Creation or the Deluge. But Enlightenment views of an active Earth, gradually changing its form over time, themselves suggested that strata and rock masses had a significant history – being formed at different times, from different materials, under different forces, and that strata could be vestiges of that history. Whereas the medals of Creation had once been fossils, now strata became the key to the Earth.

Detailed examination of the rocks – independently and within these assumptions – added body to this view. Above all the fact that strata *differed from each other* in major respects was becoming well appreciated, and was of course sparking discussion of their likely distinct formations and histories. Distinctions were being drawn between crystalline, massy and apparently sedimentary rocks; between strata which were homogeneous and those formed of a variety of materials; between mineraliferous or metalliferous, and those barren; between those fossiliferous and those without organic remains; between those which contained fossils of land, or of sea, creatures. William Hunter puzzled over giant vertebrate bones petrified in strata (1768).

Empirical awareness of such distinctions impressed even upon the minds of amateurs that Earth history was a good deal more complex and extended than had previously been imagined. When Moreton Gilks mused on the formation of strata, he used the actualistic analogy of the gradual petrifying action of limestone streams in the Peak District to reach the conclusion that the formation of strata was a much lengthier process than commonly imagined 'with regard to the received Opinion of the World's Age' (1740). Similarly, the Norfolk gentleman naturalist William Arderon sent to the Royal Society a communication in 1746 on some Norfolk cliffs. He wrote, 'The various *Strata*, which make up this long Chain of mountainous Cliffs, must be greatly entertaining to every one who takes a Pleasure in looking into the many Changes which the Earth undoubtedly has undergone since its first Creation' (1746: 276; cf. Whalley, 1971). He listed the strata, in their order, and noted that they were 'beautifully interspersed' and 'frequently the same kind many times repeated; as if at one time dry Land had been the Surface, then the Sea; after, morassy Ground: then the Sea, and so on, till these Cliffs were raised to the Height we now find them' (cf. BM Add. MS 27966, f.62–3). Alexander Catcott as well was repeatedly surprised when out in the field to find the position and disruptions of strata were far more complex than his theory, that all strata had been re-formed at the Deluge, had led him to expect. Significantly he kept such problems to himself, and in his published writings did not allow these complications to disturb the harmony of his Hutchinsonian explanations (Neve and Porter, 1977).

But up to the end of the third quarter of the century little of this accumulating knowledge of strata had yet been consolidated into tangible, usable, productive form. Predictably enough, much remained unpublished. No clear scheme had formed for organizing accruing empirical regional knowledge. Similarly, though there was a growing realization that strata were the key to Earth history, no well-defined theory of the strata was formulated: 'Nothing has more perplexed those who undertake to form theories of the earth than these appearances' truly stated the *Encyclopaedia Britannica* in 1797 (3rd ed., 'Strata'), and Earth history was only hesitantly being reconceived to take strata into account.[7]

More wide-ranging travel, greater actual familiarity with the Earth, and the new Enlightenment perspectives on Nature together focused on another feature which was to loom ever larger in the interpretation of the Earth. From the 1750s onwards, traveller–naturalists all over Europe began to fix their attention on volcanoes and earthquakes. This was in part in response to circumstances. The London earthquakes of 1750 riveted interest in England; the Lisbon earthquake of 1755 universally (Kendrick, 1956; Rousseau, 1968). Vesuvius erupted obligingly frequently in the latter half of the century. It was in part due to the vogue for Continental travel, encouraged by the Grand Tour and Enlightenment cosmopolitanism (Hibbert, 1969). Not merely were far more British travellers visiting, say, Vesuvius, but those who did were also likely to have visited Auvergne, or Cassel; Sicily or the Euganean Hills; the Giant's Causeway or Staffa, and so were apt to be impressed by the extent of volcanic phenomena, active and extinct, and their significance for Earth history (de Beer, 1962; Carozzi, 1969; Gortani, 1963).

Increased interest in volcanoes and earthquakes was itself international, including Frenchmen such as Guettard, Desmarest, Dolomieu and Faujas de St Fond; Italians like Moro and Fortis; Austrians and Germans like Baron Born and Rudolf Erich Raspe, as well as Britons like Sir William Hamilton, ambassador at Naples, Sir John Strange, consul at Venice; the famed traveller Augustus Hervey, Bishop of Derry and Earl of Bristol, and James Hutton (cf. Fothergill, 1969; 1974; Sleep, 1969; de Beer, 1962;

J. M. and V. A. Eyles, 1951). The interest was, moreover, inter-national in a very positive sense, for these naturalists forged the closest mutual contacts, and sought to produce their science on a cosmopolitan basis, just as they tried to see the Earth in the most enlarged terms of extent, time and process.

British volcanists were rich – some titled – cosmopolitan, cul-tured gentlemen of the Enlightenment. Their elite pose was as philosophers of the world, modern Ciceros or Senecas. They had contempt for ignorant, mean and narrow views of religion and Creation.[8] They admired Nature for its wisdom and grandeur. They enlarged the seventeenth-century natural theology of cosmic order, emphasizing Nature's overall permanence, activity and creativity. Christian natural theology had minimized the role of volcanoes and earthquakes in Earth history, since their associa-tions of destructiveness and imperfection threatened its theodicy. But to Enlightened volcanists these phenomena – assuredly part of Nature – *must* be wisely designed to play a necessary, construc-tive role in the Earth's economy (Hamilton, 1771: 39). Strange and Hamilton, despising the superstitious Catholic rabble who thought eruptions a Hell-fire Judgment, serenely studied vol-canoes as integral to Nature's economy. Like Erasmus Darwin or James Hutton subsequently, they underlined the stately tread of Nature's operations, the immensity of time which Vesuvius had taken to build up her successive lava flows, since these confirmed their cosmology (cf. Hutton, 1795, i: 146). Volcanoes were to be a central battlefield in the 'revolution and stasis' paradigm of Earth history. For, those who believed that active volcanoes were significantly changing the globe were more or less consciously developing an actualistic stance towards Earth history. Volcanoes realized Enlightenment dynamic natural philosophy.

The leading original theory of the age, John Whitehurst's *Inquiry* (1778), precisely embodies that mixture of old and new, imperfectly harmonized, which characterized the time. White-hurst himself typified these 'new men' becoming involved in the science (Schofield, 1963b: 144). A largely self-educated Derby clockmaker, his scientific contacts were not primarily with the traditional educated orders, but with Lunar Society entrepreneurs and local miners. Wedgwood offered to supply Whitehurst with fossils from his canal diggings in return for valuable samples of

Derbyshire spars (Schofield, 1963b: 96). His *Inquiry*, typically of provincial middle-class authors, was published by subscription. The subscribers ranged from earls to plumbers.

Whitehurst's theory began from Newton's demonstration that the Earth was an oblate spheroid (1778: 2), having gradually assumed this figure from a prior condition of undifferentiated fluidity, by the regular laws of physics and chemistry (8, ch. ii). At first, the surface was a universal ocean, inhabited by marine creatures (ch. vii). Gradually the pull of the moon via the action of the tides shaped certain land masses, 'primitive islands' (ch. iv). On these, land animals and vegetation began to thrive. All this time, however, beneath the ocean bed, potentially disruptive forces were building up, associated with volcanic activity, and emanating from some kind of central fire (ch. ix). Water, seeping through the ocean floor, was vaporized by this great heat. Steam pressures exploded the ocean bed, upon which sediments had been consolidating, in a gigantic revolution which Whitehurst associated with the Noachian Deluge (ch. xii). Such sediments were now left as the new dry land. The earlier primitive islands formed the new sea-bed after this catastrophic revolution of Nature which had caused much of the landscape derangement he now found (49). Since this revolution, the Earth had changed little.

The last third of the *Inquiry* was an Appendix, containing pictorial sections of the strata across Derbyshire and adjacent counties, with lengthy textual explanations, and some tentative generalizations about the order of the strata.

The work wove together old and new. It was still a cosmogony. It still saw the Deluge as the lynch-pin of Earth history. It still claimed harmony with the Biblical cosmopoeia and the wisdom of the Ancients (25–6). It leant heavily on Newton for its physical explanations. And Earth and human history were still intimately linked. Whitehurst's pre-diluvian Earth was a paradisical state; the Deluge rendered the Earth less fertile to compensate for man's now sinful condition (chs xiii–xvi). Yet even in these traditional respects, Whitehurst had moved beyond the seventeenth-century theorists. Though he argued that his theory harmonized with Genesis, he offered almost no analysis of Scripture (or of pagan philosophy), and his Biblical references were expressions of hope rather than of demonstration. Indeed, in some ways his theory flew in the face of Genesis, in asserting, for instance, that life was thriving in the oceans before the separation of land and sea. Similarly, though the theory was providential, there was no hint of miracles, for Whitehurst explicitly rejected all explanations beyond the present laws of Nature and actual causes. He would 'derive the nature of things from Causes truly existent' as being appropriate for an 'age which only admits of deductions from Facts, and the Laws of Nature' (p. ii).

Whitehurst owed much to contemporary science, as his citations from French naturalists like Buffon, Condarmine and Macquer witness. He stressed the regularity, gradualness and continuity of Nature's activities, both in the 'regular, uniform progression' of the formation of the present figure of the Earth from Chaos, and in the 'progressive formation' of the present strata on the sea-bed of the ante-diluvial Earth (14–17). Discreetly he kept silence on the Earth's age, but the general tenor indicates he believed it substantial.

Whitehurst was also a child of his time in explaining the Deluge not by crustal collapse or external intervention but by internal heat powers. The Deluge was deluge secondarily: its primary cause was a universal earthquake-cum-volcano, demonstrating Whitehurst's Enlightenment emphasis upon an active Earth. He also recognized that the crust testified to an Earth history which had seen revolutions of land and sea, and believed that active volcanoes gave actualistic insight into Earth history (72).

Lastly, his theory hinged on the need to explain stratification.[9] Above all, his Derbyshire evidence implied that the strata had been formed horizontally, but had undergone subsequent disruption. Furthermore, some strata seemed to have been infiltrated with volcanic toadstone at high pressure from beneath. Unique amongst theorists of the Earth, Whitehurst appended a discourse on local strata (162). Such accounts were certainly becoming more frequent, but Whitehurst's is especially detailed, being accompanied by panoramic diagrammatic sections. He stressed the utility of such for coal-finding, and ventured some low-level generalizations from his tables – such as that millstone grit is always found incumbent upon limestone but never on coal. Furthermore, he was concerned to read the meaning of fossil and strata evidence. He thus noted (170) that strata containing 'marine productions only. . . must certainly have been formed whilst the sea covered the earth. . . and those containing vegetables and no marine exuviae must have been formed after the earth became habitable. It is therefore apparently repugnant to the general course of Nature that terrestrial animals and vegetables should be blended together with marine productions in the primary strata of limestone' (170). The *Inquiry* was thus a milestone in the better integration of field observation with theory, made possible precisely because of growing recognition of the strata.

Yet Whitehurst's theory also reveals the shortcomings of the age. The fact that the *Inquiry* preserved a formal distinction between a general theory and an entirely separate Appendix of local facts speaks for itself of continuing tensions between traditions of theorizing and Baconian observation. Furthermore, the data set out in the Appendix frequently flew in the face of the theory. Whitehurst listed, for example, in the Appendix a vertical sequence of barren and fossiliferous strata, some of which contained terrestrial and some marine exuviae. But his theory had no mechanism to explain these facts (170f.) – nor did it address itself to the issues involved. Likewise, the theory offered no discussion of the mechanism of strata formation or why distinct strata, containing different materials, should have consolidated. Fact and theory made ill bedfellows.

Up to the end of the third quarter of the century, new trends showed more promise than achievement. Much work on strata remained unpublished – and unpublishable. Even the prolific Sir William Hamilton desisted from organizing his broad knowledge into a work more comprehensive in explanatory ambition than empirical accounts of Italian volcanoes (1772; 1776–9). The period contains hints, rather than treatises, about the age of the Earth, its processes, the philosophy of stratification. And theories such as Whitehurst's were patently flawed. The reconstitution of Earth science was still left as a task for the next generation.

# 6

## Changing social formations: *c.* 1775–1815

We have seen so far that investigation of the Earth, to about 1710, boasted a socially coherent group of devotees – the traditional educated elite – and a compact body of ideas. Though empirical investigation of the Earth was novel, it became organized within disciplines – cosmogony and natural history – that were blessed with respectable pedigrees. The movement faltered when theories of the Earth produced intellectual anarchy not unity; ushering in a period of disaggregation through the central decades of the eighteenth century, a time of modest experiments with new forms of scientific organization, techniques and visions: a period symbolized by that ubiquitous Enlightenment figure, the traveller (Carswell, 1950). This was a formative age, though scarcely one of assured accomplishment. Activity was decentralized and was often hampered by intellectual isolation, and fumblingly premature attempts by new kinds of naturalists to create new forms of Earth science. The profound stirrings of mid-century – the growth of an audience, the birth of provincial science, the spread of field-work – had not yet borne fruit in coherent research programmes or powerful theories. The time saw awareness of the need for reformulations, rather than the achieved reformulations themselves.

In the last quarter of the century many naturalists continued to feel isolated, and to ponder problems of method and theory. But a greater assurance was crystallizing of how investigation of the Earth was to be realized in scientific form. This confidence was reinforced by consciousness of the science's dramatic explosion. John Hailstone, Woodwardian professor in Cambridge, captured this in his first lecture. Noting that none of his predecessors had actually lectured, he remarked:

I mean not in saying this to cast the smallest reflection on the conduct of my Predecessors. Till within these few years the general attention of mankind was directed in Natural History to the departments of Botany and Zoology. The sister branch of Minerals seems to have been almost totally overlooked. It consisted of a few scattered unconnected facts, incapable of being digested into a system. And what is incapable of being reduced to a system cannot be made the subject of public instruction. The rapid progress which has since been made in Mineralogical knowledge has however made ample amends for the neglect with which it was first treated.[1]

The remaining chapters of this book will examine the quantitative growth of Earth science.

But I shall also investigate what Hailstone and many others meant in affirming that at long last the science had got an adequate 'system' as a foundation for further labours. I shall argue that scientific study of the Earth became *geology* in this period. This is an important semantic point, in that the word 'geology' first came into common use in the last quarter of the century (Adams, 1933a; 1933b). But it is far more.

Some historians, such as C. C. Gillispie (1960: 299) and Leonard Wilson (1972), have contended that geology did not become scientific until 1830 with the first volume of Lyell's *Principles*. Before Lyell – the simple argument would run – there were merely *a priori* speculation and religious rhapsodies about the Earth. I have strenuously argued that *this* certainly is not true. On a more sophisticated level, it is claimed that Lyell drew out of chaos a new framework for conceptualizing the Earth, and a new method, Uniformitarianism, both of which became paradigmatic. But this interpretation is weak in rear and van. For, many of Lyell's immediate predecessors and contemporaries themselves already and independently held the more plausible tenets of his world-view and method. Furthermore, it has now become clear that Lyell's *Principles* did not in fact command paradigmatic assent (Cannon, 1960a; 1960b; 1961).

Much more convincingly, it has been argued – above all, by Martin Rudwick (1971) – that by the 1820s there was growing consensus upon a general view of Earth history, and a programme and method of research – 'directionalism'. Yet, many of the components of this 'directionalist' paradigm, such as the emphasis upon the sequence of strata and life, a long time-scale, naturalistic explanation, the progressive development of the Earth, were continuous with – not reactions against – interpretations which had been steadily crystallizing during the last third of the eighteenth century. Furthermore, to say that a closely-knit community of scientists in the 1820s held a 'geological' paradigm presupposes the social existence of that community. But that is something which had had no defined and tangible reality in Britain in, say, 1770. By the 1820s, however, it was a real entity which marked itself off, for example, by membership of specialist societies.

The last chapters of this book will therefore seek to trace the forming of this community and its geological outlook. The community was to possess a social coherence similar to that of the late seventeenth century; but it had a vital depth hitherto unknown. Its beliefs also possessed great coherence, but the earlier plurality of scientific approaches now additionally produced, in geology, a new science which was to outstrip all in appeal and intellectual power.

The precise moment of the emergence of characteristically English geology at the turn of the nineteenth century was relatively sudden, the product of a conjunction of internal and external causes, changes relatively necessary, and causes relatively contingent to the science itself. The ordering of the next three chapters attempts to 'unpack' the different layers of events, each with its own chronology and time-pulse, and then to reintegrate them.

This chapter seeks to locate the English social and institutional milieu which generated and supported an Earth science which became geological. By way of a social control, I then examine the very different social, institutional and intellectual context of Scotland, to try to establish that, though producing extremely eminent Earth science, Scotland did not generate geology in the same mould as in England. In chapter 7 I seek to show the *deep* and *general* intellectual transformations, relatively internal to the practice and problems of different aspects of Earth science, which were requiring them to be restructured within that overarching vision of the Earth which we call geological. The last chapter then shows how that general and slow transition to geology was overtaken and deeply affected by contingent forces. The first was the appearance (in many respects, from without) of a wave of theories of the Earth in the 1780s and 1790s. Though more sophisticated than previous ones, theories such as those of Hutton, Toulmin, Erasmus Darwin, de Luc and Kirwan were extraneous to the empirical temper of the age, and redoubled the desire to define geology over and against theorizing. Furthermore, at precisely this time, the deluge of religious and political counter-revolutionary feeling provoked by the French Revolution served as society's notice to geologists as to the limits of any currently acceptable Earth science. The geology gradually forming under deep social and intellectual stimuli was brusquely forced into an extreme empiricist posture by the exigencies of the times. But this actually proved a recipe for success. Such extreme empiricism was invaluable in generating the self-sustaining nature of geological science in the early nineteenth century.

## Professionals and amateurs

British society was experiencing unparalleled expansion in all directions in the late eighteenth and early nineteenth century. Within it, the networks studying the Earth were no exception. Sir Joseph Banks, for instance, in his own right and as president of the Royal Society, corresponded with many hundreds on mineral subjects (Dawson, 1958). Soon after the founding of the Geological Society of London (1807), certain members were concerned that too many wished to join (Rudwick, 1963: 337). It is attractive to seek to explain this growth in Earth science in terms of industrialization, seen as both an economic and a cultural phenomenon. For industrialization could *prima facie* quite directly stimulate geology, by producing needs within mining for geological expertise, and by giving urgency to understanding and exploiting the nation's natural resources.

Discrimination should be exercised, however, in ascribing the making of geology to mainstream economic causes – even in those sectors where direct industrial needs and pressures apparently stand out. Even where the Industrial Revolution was creating unprecedented needs for knowledge of the Earth, and potential capacity for supporting practical geology experts, the very *laissez-faire* capitalist structure of that industrializing process was simultaneously militating against creating these experts and that practical science. Thus even the putatively more direct economic causes are complex and fully mediated. I have argued elsewhere that despite the *a priori* plausibility of linking the Industrial Revolution and the emergence of geology; despite contemporary hopes and expectations that there should be intimate relations between geology and industry, especially mining; and despite the intimate connections which apparently *did* exist between industry and certain sciences (e.g. chemistry); any immediate relationship between industrialization as such and the making of geology is more problematic (Porter, 1973b). Very few who possessed intimate professional knowledge of the Earth made significant contributions to the development of geological concepts or theory – whereas those who did owed little directly to personal contact with surveyors, miners, or with mining. The leading early nineteenth-century geologists were only marginally involved in the practical deployment of their knowledge.

That having been said, it would be gratuitous not to point out the substantial common ground between practical men and geological scientists. No geologist would spurn a chance to go down a mine or talk to a miner. Equally, men who like William Smith made a living out of geological expertise certainly did not feel obliged to desist from theorizing (L. R. Cox, 1942–5: 84–5). There was a broad interface in late eighteenth and early nineteenth-century Britain between miners, surveyors, consultants, and other practical men on the one hand, and the geological intelligentsia on the other (Bush, 1974: ch. iii). This is no less true, even though the actual relations between them were not as cordial, as forthcoming, or as fruitful as many would have liked; and even though it later became clear to most observers, and even to government, that in this science, as in most, greater contact between science and industry was vital for the future (Cardwell, 1972).

The late seventeenth-century group of investigators had owed its cohesion to a common background in Oxford and Cambridge education. A century later, the network was far more socially heterogeneous. It is true that the great majority were linked by being middle class in origin and life-style – which had the consequence of creating social tensions *vis-à-vis* men of humbler social position such as William Smith. But the late Hanoverian bourgeoisie was itself so heterogeneous – in wealth, occupations, professions, in education, in religion – as to compel us to look beyond its social make-up for the closeness of the geological community. The answer is that, for the first time, it was the science itself which achieved sufficient cohesion, attraction and importance as to be its own social cement.

This is apparent if we examine two institutions which brought together a cluster of geologists, as a contrast with earlier societies. The members of late seventeenth-century societies had been united by common background at least as much as by common purpose. In the case of the Dublin Philosophical Society, they were the educated Irish gentry; in the Oxford Philosophical Society, they were M.A.s of the University. Both societies lacked the size, social depth and *esprit de corps* which would have given them more than fleeting existences.

The Literary and Philosophical Society of Newcastle-upon-Tyne, founded in 1793, was part of the blossoming of scientific societies in industrializing England discussed earlier.[2] Though similar in aim to the Lunar Society, it was differently organized. For, while the Lunar Society was an *ad hoc* private group of friends, the Newcastle society was a formal institution open to all residents (Watson, 1897: 329–30), on the model of the Manchester Literary and Philosophical Society (Thackray, 1974). But whereas the Manchester society had little local reason to cultivate geology, the proximity of Britain's most economically important coalfield and valuable lead workings ensured that the interface between geology and mining would be fundamental to the Newcastle society.

Indeed, from the beginning that was the intention. During its first thirty years, the issues of practical local geology and of mines safety – obviously mutually connected – constituted perhaps the two most unifying strands in the society's activities. Members came to hear papers on local geology; paid for Robert Bakewell's courses of geological and mineralogical lectures, and financed the upkeep of extensive mineralogical collections – praised by Bakewell (1815: 88) and Hatchett (1967: 78); and some personally wrote geological essays. The society's programme promised the parallel advancement of rational knowledge and local wealth, through co-operative local scientific inquiry (W. Turner, 1802). The enterprise of a practical geology of the north-east was able for many years – years of general political tension – to enlist substantial support from the local community, from Unitarian and Anglican clergymen, landowners and coal-mine owners, through tradesmen, doctors, journalists and booksellers, down the social scale to coal-viewers. And the aspirations and support continued despite the failure of the society collectively to achieve tangible results, above all in its scheme for a composite geological map of the mining area.

The Newcastle Lit. and Phil. was the first provincial society to plant its hopes centrally on the scientific–economic connection between geology and mining. Many such societies – both Lit. and Phils. and provincial geological societies – were founded subsequently in the nineteenth century, providing a lasting focus of local research in the science, and a recruiting ground for amateurs. Several of the papers presented to the Newcastle society were subsequently published – some independently, some in the *Transactions* of the Geological Society, and some in other periodicals such as the *Monthly magazine*. Its members' geological expertise was tapped – or purloined – by visiting geologists. Perhaps because of the Newcastle society, the geology of the north-east was unravelled largely by local men, rather than by the national elite of peripatetic geologists.

The Royal Geological Society of Cornwall (f. 1814) provides a similar example.[3] In the eighteenth century, Cornwall had not even supported a *general* scientific society. Three main factors brought the Cornwall geological society into existence. Firstly, geology had become a coherent pursuit by 1814. Tin and copper mining in the county had become precarious, with mines sunk ever deeper and becoming more capital-intensive (Burt, 1969). Thirdly, some of the Cornish mine-owners had gained personal experience at the Freiberg Mines Academy and other Continental mining schools, in which geology, mineralogy and metallurgy were all promoted alongside the furtherance of mining skills (Hawkins, 1818; Vivian, 1815a; 1815b; 1818). Like the Newcastle Lit. and Phil., the Cornwall society was remarkably successful in attracting powerful support from a wide social spectrum, which included several aristocrats (with the Prince Regent as patron), substantial gentry, mine-owners and lessees, clergy, doctors, lawyers and a large number practically involved in mines working. In effect the community who had been prepared to subscribe to Pryce's *Mineralogia Cornubiensis* in the 1770s were now joining together in a geological society.

Despite clashes over policy, the Royal Geological Society of Cornwall (Todd, 1955–64: 181) responded to the needs and wishes of its local community by pursuing investigations into mining, mineralogy, metallurgy and mines safety as well as pure geology – and most important, by pursuing the *relations* between geological knowledge and its applications with more success and conviction than the Geological Society of London. Whereas the latter paid lip-service to utility (cf. *Geological inquiries* (1808), and the Preface to its *Transactions*, 1811; Porter, 1973b), the first two decades of articles in the Cornish *Transactions* show some twenty-seven mining articles, about twenty on geology, some eighteen on mineralogy and about six on metallurgical subjects.

The success of the Newcastle and Cornwall societies illustrates how a range of enthusiasts, highly disparate in terms of rank, class and profession, could be brought together by the end of the eighteenth century as a result of a shared concern with geology, within local communities with powerful mining interests. By the early nineteenth century provincial regions could thus support geological societies. That mining interests underlay the societies – not dilettante gentlemen – ensured that the interest in the Earth would be geological – focusing on strata – not cosmogonical or natural historical. Thus the secretary of the Newcastle Lit. and Phil., William Turner, wrote that the society should pursue geology 'not as employed in speculating on the original formation and structure of the Earth, but as having for the object of its inquiry the relative situation of its productions as Nature now presents them to our attention' (1802).

In addition to local communities whose involvement with the Earth was largely a function of regional economic interests, two social groups were rising in numbers and social prestige towards the end of the century, who were not only to contribute very substantially to Earth science over the next generations, but whose social and intellectual perspectives directed that science towards geology. These were professional surveyors and mines consultants on the one hand, and, on the other, affluent, lay members of the liberal professions, including professional scientists, usually living in the metropolis.

Essentially for the first time, those whose professions demanded extensive knowledge of the Earth – land surveyors, canal builders, coal prospectors and viewers, drainage experts, mineral assayers, quarrymen and civil engineers – were seeking to elevate their professional expertise into 'scientific knowledge' (Pollard, 1965; Duckham, 1970: chs. v, ix; Bush, 1974). The absolute size of such professions was increasing. Their professional status was rising within an industrializing society. Possibly, men of greater talents were being attracted into them. Such trades had always possessed traditional Earth lore. But not until quite late in the century was such knowledge commonly publicly printed. Perhaps engineers had chosen not to publicize their knowledge before, in order to keep craft techniques secret. Possibly they had been in no social, educational or financial position to do so. By the end of the century, however, such men successfully strove to transform craft wisdom into scientific knowledge of the Earth. Among the most important were John Smeaton, John Williams, William Smith, John Farey, Robert Bakewell, White Watson, Westgarth Forster, Benjamin Bevan, David Mushet, Edward Martin and Benjamin Outram, but many more could be mentioned (cf. Bush, 1974). In part, their scientific ambition was professionally self-interested. Public proof of skills, evident in publications, could be good for business. Publications could make money. But their writings also communicate scientific interests moved by a general patriotic Baconian spirit of the advancement of knowledge: 'The PRACTICAL ADVANTAGES of Geology, and its beneficial application to Agriculture, Mining, Coalworking and Commerce' as William Smith put it (Phillips, 1844: 109).

Many laboured under great difficulties to produce scientific work. Their formal education was often minimal.[4] Few had any deep grounding in general natural philosophy beyond their specific fields, or could read foreign languages. Leonard Horner wrote of Farey, 'He seems to proceed as a geologist without the aid of science' (1811b). Their professional work left little leisure for general reading, or literary composition. They felt themselves cut off from libraries, and from the scientific patronage of the metropolitan intelligentsia. They accused fashionable society of failure to support their work or to subscribe to projected books. They complained of lack of government-sponsored posts for technical experts. They deplored society's neglect of useful sciences, such as stratigraphy.

In addition, these surveyors were frequently competing against each other for jobs and fees. Hence mutual co-operation on scientific work was sporadic and unstable. Bakewell had little contact with Smith. Smith and Farey, initially friends, quarrelled. Farey and Bakewell had a war of words in the *Philosophical magazine*. And, above all, they complained of financial difficulties, which at the least curtailed and delayed projects, led them to publish in the cheapest form possible, with little regard for the most effective presentation, and which, especially in the case of William Smith, were a stumbling block to ambitious undertakings (Phillips, 1844: 47; Ford, 1960). Summing up Smith's social situation, John Phillips made a Smilesian virtue out of necessity: 'Mr Smith was utterly unacquainted with books treating of the natural history of the earth; he had no other teacher than that acquired "habit of observation" which he has so justly recommended to his followers. It is difficult in these days to conceive of such insulated and independent research, as that into which the young philosopher entered' (Phillips, 1839: 215; though see Bush, 1974: ch. i).

Such handicaps left their mark in many shortcomings of their work. Their theories were often dated or jejune. William Smith in particular seems to have cut himself off during the last thirty years of his working life from conceptual developments in the science. Some of his statements about the order and above all the formation of strata betray painful ignorance of contemporary discoveries. At one stage Farey accepted a theory remarkably like Whiston's (1807; 1811). Many of Farey's and Bakewell's periodical articles are higgledy-piggledy collections of observations, strung together without interpretative form (Farey, 1814; 1815a). Their stratigraphical work was often excessively localized. Even Farey's monumental *Derbyshire* breaks down (e.g. in interpretation of faulting) because of ignorance of the adjacent counties (1811–15; 1815a). Ford has pointed out that there is no evidence that White Watson ever travelled more than twenty-five miles from his native Bakewell (1960: 350).

Yet such men frequently had unmatched empirical knowledge of one salient department: the strata. Canal construction revealed the continuity of outcrops over considerable lengths of country-side, and particularly their effect on topography (J. M. Eyles, 1969). It demanded knowledge of the relative porosities of strata, of springs and water-courses. Agricultural surveying showed the relation of strata to soil. Coal mining laid bare the vertical succession of the strata, and prospecting required an ability to extrapolate from such knowledge.

The social and professional experience of these 'new men', then, directed their Earth science in two main respects. Firstly, they focused on the strata (hence 'Strata Smith'). Their work programmatically emphasized that stratigraphical knowledge was the key to understanding the Earth (Townson, 1797; Phillips, 1844: 107–8). They attacked preoccupation with fossils, minerals and theories taken as ends in themselves. They ignored the analysis of landforms and Earth processes (Davies, 1969: 204). Their work gave weight to the central role of stratigraphy within emergent geology. Secondly, they derived their success from personal effort, and their knowledge from personal investigation. This led them to champion meticulous observations, and to publicize the more populist features of Baconianism: hostility to speculations and verbal science, and confidence in practical men's role in scientific progress (L. R. Cox, 1942–5: 85f.). Such utilitarian vulgar Baconianism was annexed to become the universal ideology of early nineteenth-century geology, however little it actually reflected the practice of most elite geologists (Porter, 1973b).

No less important in the forming of geology was the contemporary, dramatic surge of interest among higher ranks in society – well educated, comfortably placed members of the more liberal professions, and gentlemen of leisure. Of course, men broadly of this social type had been involved from the mid-seventeenth century. But important shifts are evident. Firstly, as already discussed, an indigenous middle-class *provincial* scientific culture had grown up from nothing (Thackray, 1974; Shapin, 1971). Secondly, the range participating in 'establishment' science in London was spreading. Only a small minority of the early active members of the British Mineralogical Society

(f. 1799) and the Geological Society of London had had an Oxbridge education. Many more had been educated either at a Scottish university, at a Dissenting Academy, privately, or abroad. Some were Anglican in religious affiliation, but varieties of Dissent were to the fore, with Quakers, such as William and Richard Phillips, William Pepys and William Allen, pre-eminent among the founder members of both societies (Raistrick, 1968). Unlike the preponderance of seventeenth-century devotees, none of the founder members of the Geological Society of London was in Anglican Orders. Furthermore, the range of employment and source of income had widened. Most of the late seventeenth-century community had had academic or ecclesiastical posts. But the London geologists at the turn of the nineteenth century also include publishers, professional chemists and other men of science, businessmen and bankers, as well as those in trades connected with the practice of science – such as draughtsmen and engravers.[5] The group was notably non-academic and non-clerical. It was broader in base than before, which gave it greater solidity. It was held together explicitly by commitment to geology.

The other salient feature about the founder members is their tendency towards scientific specialization. Specialization is relative. To our eyes, Arthur Aikin might seem a polymath rather than a specialist, since, as well as following geology, he was among other pursuits briefly a Unitarian minister, an editor, an encyclopaedist, a practising chemist, and the secretary of the Society of Arts. But his avocations were narrow compared with those of his father, John Aikin, who had been a humanist and theologian, as well as a scientist. Arthur was essentially a scientist.

Compared to the breadth of religious, humanist and scientific interests of Lhwyd, Ray, Lister and Whiston a century before, or to the range of sciences in which Hooke, Halley and Woodward excelled, the age of Charles Hatchett, Arthur Aikin, Leonard Horner, William Babington, William and Richard Phillips, George Bellas Greenough, J. J. and W. D. Conybeare and John MacCulloch is one of narrowing specialists. This trend towards specialization is visible in other contemporary forms connected with the emergence of geology, such as the appearance of distinctively scientific periodicals like Nicholson's *Journal*, or the foundation of specialist scientific societies. When in 1796 a group of young London men of science sought an alternative to the Royal Society, they first set up a general science society, the Askesian. Out of this developed more specialized bodies, one of which was the British Mineralogical Society (f. 1799), several of whose members became founder members of the Geological Society of London (H. B. Woodward, 1907: 6–14). Division of labour and specialization arguably led the members of the specialist institution to cohere more tightly. The community of geologists in the early nineteenth century was bound together much more firmly than any equivalent group a century, or even half a century, earlier.

How do we explain this specialization? It is certainly not coterminous with professionalization. Babington, Greenough, Horner, W. D. Conybeare, William Phillips and many others involved themselves in geology almost to the exclusion of other scientific interests, but essentially as amateurs. Not until the establishment of the Geological Survey in 1835 did an Englishman of any rank make a regular and secure living out of his skills at geological fieldwork (Bailey, 1952: 21). Specialization in Earth science was in fact part of a general movement towards specialization in all sciences, given a strong impulse by the astounding rate of increase in available information in the early nineteenth century which made involvement in a wide range of sciences increasingly impractical (Cannon, 1964b). Specialization in geology was perhaps aided by fading belief in the Great Chain of Being (Bynum, 1975), thereby sanctioning investigation of the mineral kingdom in isolation from the other two, and by strictures against geo-physical cosmogony, which permitted study of the Earth independently of the exact and physico-chemical sciences.

But why should there have been a surge of interest in *science* within metropolitan society at the end of the eighteenth century? Why did much of it focus upon the Earth, taking the form of geology? The geological intelligentsia at the turn of the nineteenth century was a cohort of young men. A radical current of idealistic earnestness, a commitment to enthusiasm and activity was permeating the intellectual young (cf. Meteyard, 1871). With some, most obviously Dissenters of the kind who comprised the Askesian

Society, the British Mineralogical Society and the founding members of the Geological Society, exclusion from the political and religious Establishment fostered zealous commitment to the progress of utility, wealth, knowledge and talent (Raistrick, 1968). For others, the mood registered idealistic frustration at, and rejection of, corrupt late eighteenth-century aristocratic polite society, focused by defeat in the American War and the movement for political reform. It also, clearly, marked personal ambition. The Enlightenment ideal of cosmopolitan stoicism gave way to passionate Romantic engagement – witness the precipitant energy of the geological activities of the young Thomas Beddoes, Edward Daniel Clarke, Humphry Davy, Sir James Hall, with his hair-raising traverses of the Alps, or William Buckland's headlong horse-riding across Britain. Characteristic was the ardour of the affectionate and infectious friendships which many of this network shared; their spritely personal letters; their fieldworking in little groups while holidaying together. They created and experienced the excitement of young men accomplishing something novel and daring. 'I partake more largely of the spirit of the Knight of La Mancha than of his craven squire and prefer the enterprise and adventure of geological errantry to rich castles and luxurious entertainments' (W. D. Conybeare, 1811).

This radical energy determined their choices. Dissatisfaction with traditional polite society led many, particularly Dissenters, to the pursuit of science. Natural truths were worthier than social conventions. Science was useful, progressive and of national importance. Of Benjamin Richardson, his obituarist wrote that his early feeling 'of the higher value of sciences when compared with polite literature never deserted him' (W. S. Mitchell, 1870–3: 325). Chemistry – both practical and theoretical – in particular attracted huge interest during these years, stimulated by the French revolution in chemical thought. Many young chemists who took up geology had been early converts to French chemistry – for example Sir James Hall and Thomas Beddoes.

But why should the science of the *Earth* be so attractive? One stimulus was travelling. The surging popularity of scientific travelling as recreation or as Grand Tour merged insensibly, with men such as Hall, Greenough, Clarke, and Buckland, into a more Romantic form of travel as *Wanderlust* – travel as a metaphor for education and experience. For Edward Daniel Clarke, 'unbounded love of travel influenced me at a very early period of my life. It was conceived in infancy and I shall carry it with me to the grave' (Otter, 1825, i: 123).[6] Travel and science fused in the painstaking observation of the Earth.

Travel was fed by Romantic sensibilities (Knight, 1970; Treneer, 1963). Taking up science against allegedly stifling polite culture they sought Nature in her wild and unspoilt forms. Arthur Aikin (1797) and William Maton (1797), amongst others, toured through the more scenically dramatic parts of the British Isles to produce accounts of the picturesque amongst the mineral structure. Davy was not alone in seeking solace and knowledge together in the rocks (Treneer, 1963: 80–1). Romanticism broadly encouraged exploration of the natural environment. But it also gave specific shape to Earth science. Travel encouraged fieldwork over an extended terrain, which focused attention on the strata. Romanticism extended Enlightenment projection of the powers, activity, unity and grandeur of Nature. To the Romantic mind the preoccupation with identifying isolated specimens, which typified long-established natural historical and mineralogical studies, was artificial and petty-minded. It put human systems before Nature's realities. The Romantic mind wished to take the Earth as it was in Nature, not in the laboratory. Geology was to be precisely that kind of knowledge. The geological vision hinged upon the belief that to read the story of the strata was to read an autobiography of great revolutions, decay and restoration, the struggle of titanic Earth forces. By the early nineteenth century, for practitioners and for the general public alike, geology was assuming the image of a more popular, profound and *spiritual* science than mineralogy or orthodox natural history (Rudwick, 1963: 328; Cannon, 1964a; Nicolson, 1959: 371f.).

Thus investigation of the Earth by this elite grew more specialized and demarcated from other fields of science. The work of the *lay* middle class, it became distinguished from overtly religious forms of Earth science. It combined a traveller–naturalist's interest in strata with a Romantic concern with the complex, lengthy, active development of the Earth. The investigations of this elite took a generally similar direction to those of the practical surveying tradition already discussed. Of course, the two groups were not rigidly segregated from each other. They had much contact, though of a slightly tense form, involving deference, patronage and suspicion. A minor class war between certain members of the different groups flared in the 1810s.[7] It was fuelled by misunderstandings and personal temperaments, but these were necessarily latent where groups were competing for priority rights in colonizing a new scientific territory (Porter, 1973b: 325–6). By the early nineteenth century, investigation of the Earth had an assured, expanding social basis. One band was committed to the science by professional standing. Another cluster, through its own cultural choices, had made the pursuit of geology the expression of its own commitment to intellectual work, Nature and the radical consequences of Truth.

It would be a mistake to look to a very high degree of social and institutional homogeneity as the condition for the success of geology. Perhaps no science has ever achieved the solid institutional unity presupposed by certain sociologists. Rather British geology in the early nineteenth century was multifarious and differentiated in its social bases – between metropolis and provinces, practical men and the wealthier elite, Scotland and England – with style and content varying amongst the different schools. But diversity and conflict were probably the ideal environmental stimuli for the phenomenal growth of British geology between 1790 and 1830.

## Geology and its institutions

Earth science retained its place in established general scientific institutions, both formal and informal, at the end of the eighteenth century. For example, the Royal Society, despite the strictures of the Geological Society of London, published a fundamentally important series of papers on volcanism, and the first British papers on vertebrate palaeontology. Its president, Sir Joseph Banks, pioneer explorer of Staffa, maintained a huge international correspondence on mineral subjects. Similarly *new* general scientific media and societies were giving a major place to Earth science. Popular scientific journals such as the *Philosophical magazine* (f. 1798) and the *Annals of philosophy* (f. 1813) gave prominence to geology. From 1797 the newly formed Edinburgh

Academy of Physics singled out the theory of the Earth as one of three scientific problems demanding urgent solution (Cantor, 1974).

Existing specialized Earth science institutions now began to flourish as never before. John Hailstone, the sixth Woodwardian professor at Cambridge, became the first to lecture. Hailstone himself had studied under Werner at Freiberg, and his printed syllabus of lectures had a thoroughly Wernerian content (1792). He actively geologized in Cornwall and Scotland (Rudwick, 1962: 128). Hailstone's lectures were supplemented from 1809 by Edward Daniel Clarke, who was granted a personal chair in Mineralogy (Otter, 1825: ch. viii). His first lecture was attended by two hundred. In Oxford, William Thomson and Thomas Beddoes in the 1780s, and then a little later John Kidd and William Buckland, began a continuous tradition of teaching mineralogy and geology (Waterston, 1965; Stock, 1811). In Edinburgh, John Walker, Thomas Charles Hope and Robert Jameson established geological and mineralogical lecturing from the 1780s (Scott, 1966). In Dublin, Carl Giesecke lectured (J. M. Eyles, 1961). Thus by the end of the century geological lecturing was regularly established in the ancient British universities. Such growth, though unsurprising, was significant. The rise of geological teaching institutionalized the science as an option within the educational system. In future, geology was less likely to be a passion developed *despite* one's education. Charles Lyell is a notable early example of one who, having attended geological lectures in Oxford, himself became an ardent geologist (Wilson, 1972: 42).

Old-established collections also expanded. The British Museum collection was augmented and arranged by Charles Koenig (W. C. Smith, 1969). Robert Jameson turned the Edinburgh University Museum into a superb teaching collection (Chitnis, 1970; Morrell, 1972–3: 49f.). The Leskean Collection of minerals, amongst the finest in the world, was bought for the Dublin Society (Sweet, 1967). Great private collections flourished – those of the Earl of Bute, Charles Greville, Sir John St Aubyn and Philip Rashleigh. Mineral dealers and semi-professional fossil-gatherers began to spring up (Lang, 1935; Ford, 1960; Murray, 1904: ch. xii).

Two significant developments secured the growth of geology as an independent specialist science. The first was the major role played by geology in newly-founded societies. I have already discussed the importance of geology in provincial scientific societies oriented towards utility, like the Literary and Philosophical Society of Newcastle and the Royal Geological Society of Cornwall. In London, though its elaborate plans for establishing mineralogical collections and laboratories came to little, the Royal Institution's geological courses by Humphry Davy and W. T. Brande were highly successful (*Phil. mag.*, 1804: 284; Brande, 1816). In Dublin, the Royal Irish Academy gave prominent attention to geology, with its president, Richard Kirwan, in particular contributing a notable series of papers to its *Transactions* on subjects ranging from Irish landforms to Huttonianism (1802a; 1802b).

The Royal Society of Edinburgh, however, was the most important new general scientific society in which geology became a leading activity (Shapin, 1971; 1974). Before its foundation in 1783, very little Scottish Earth science had been published, despite the existence of the Philosophical Society. But on the establishment of the R.S.E., Hutton almost immediately wrote up part of his theory to deliver to it (Playfair, 1805: 50–1). Hutton and some of his closest geological friends – Joseph Black, John Playfair, Sir James Hall, Thomas Hope, and then later Sir George Mac-Kenzie – became key members, using the society as a vehicle for their geological views. Between 1807 and 1811 geological papers were almost outnumbering all other contributions put together. Geological papers were even read in the Literary Class of the Society (R.S.E. Minute Book, 18 June 1798). The society served as a theatre for the Huttonian–Wernerian debate, and the unprecedented step was taken in 1811 of setting up a standing Geological Committee because geology was of such controversial importance (Minute Book, 20 May 1811). Even after the Wernerian group, led by Jameson, in effect seceded in 1808 to form the Wernerian Natural History Society, the Royal Society of Edinburgh continued to debate geological issues more regularly, and seemingly with greater gusto, than any other science. In short, then, by the close of the eighteenth century a rich potential of geological thought was being tapped within newly established societies.

But the founding of the Geological Society of London in 1807 was the major landmark of the making of geology as an independent science. Since its origin, constitution and activities have been analysed elsewhere (H. B. Woodward, 1907; Rudwick, 1963), I shall merely pinpoint its significance within my theme.

That it was an *independent* geological society is crucial. Its fraught struggle for independence from the Royal Society of course spotlights the autocracy of Sir Joseph Banks and the eager ambition of the geological Young Turks (Rudwick, 1963: 341). But it also marks the moment at which a sufficient geological community had gelled to wish to organize itself independently rather than to be part of a society which spanned and purported to integrate all the sciences. Geology was now, they thought, a self-contained discipline, which could properly be pursued in relative isolation from other sciences.

Secondly, though as Greenough wrote its founding 'was the effect of accident rather than design', it is pertinent that it became a *geological* society. Many of its founders had been members of the short-lived British Mineralogical Society (1799–1806).[8] Most of them began as chemists or mineralogists, not as geologists, and first met to arrange translation of a work by the French mineralogist and crystallographer, the Count de Bournon. However, the British Mineralogical Society had failed. It had set itself the utilitarian but *passive* task of analysing specimens sent up casually from the country. Few specimens were sent, and interest tailed off. Hence, by 1807, an *active* geological programme, of personally investigating Nature, clearly appeared more attractive and promising. Certainly, some influential members believed that a *geological* society would have more public appeal, as a more comprehensive, elevating and less dry science. The bug of geological travelling – seeing Nature 'on a grand scale' – had now bitten Horner, Aikin and Greenough (Rudwick, 1963: 328).

In its Reformation break from the Royal Society, the Geological Society went back to the pure gospel of Baconian empiricism. Its major members' private letters, and its public communications, such as its questionnaire the *Geological inquiries*, trumpet Baconian empiricism loud and clear (cf. Lyell, 1830: 59). The Society's aim was first to collect facts rather than to form theories: to stimulate and integrate research collected from diverse sources into a coherent whole; to publish knowledge; to build up a bank of data, collections and specimens.

This was a formula for success. By giving priority to further research, the Society served its members with a programme of future work; agenda for discussion; subject matter for publication. By indefinitely postponing the millennium of a theory of the Earth, it thereby guaranteed its own future. By ruling out premature discussion of theories, it sought to avoid divisive factions amongst its members of the kind which broke up the Royal Society of Edinburgh into smaller rivals. But by emphasizing that the Society's business was to put empirical research into *order* – by defining terminology and nomenclature, drawing up maps, building collections on distinct classificatory bases – this ideology gave the Society a rationale as 'a centre where mineralogical facts might be accumulated and properly arranged' (Warburton, 1814), for directing and orchestrating geological work undertaken throughout the country.

There are mildly patronizing overtones in the claim of the Society's *Geological inquiries* (1808), that

to reduce Geology to a system demands a total devotion of time, and an acquaintance with almost every branch of experimental and general Science and can be performed only by philosophers; but the facts necessary to this great end may be collected without much labour, and by persons attached to various pursuits and occupations; the principal requisites being minute observation and faithful record. (reprinted in *Phil. mag.*, 1817: 421)

But precisely that confident, even arrogant, elitism which the Society exuded in its early years contributed to its success. It ensured its independence from the Royal Society, helped to enlarge its membership amongst the influential, and to impose its vision upon the English pursuit of geology. Despite professions of integrating the labours of artisans and miners within the geological community, and of aligning geology to promote mining, the Geological Society of London was, from the beginning, a society for gentlemen amateurs (Rudwick, 1963: 353f.; Porter, 1973b: 322–3), unbendingly pursuing the science on their own terms. The core of the Society was a group of affluent friends, who could dine together and be clubbable on intimate terms. When it sought aid from the provinces, it turned as much to the landed gentry as to surveyors and mines managers. While fieldworking, the likes of

Buckland and Conybeare obviously enjoyed cutting a figure as
slightly eccentric gentlemen who could command the local quarrymen and molecatchers to help them on their business.

The Society set about imposing its vision with great confidence.
It formed committees to decide the pressing problems of universal
nomenclatures and classifications, to draw up maps, to settle a
basic vocabulary for the science. It set up series of elementary
lectures to instruct beginner members in the rudiments. With the
aid of many members and a committee of the Society, Greenough
went about constructing a definitive geological map of England
and Wales as the apex of the Society's endeavours (Bush, 1974:
ch. vii). The Society's *Geological inquiries* set out a programme
for research. And, not least, the format and scope of the articles
published in the early numbers of the *Transactions* were standardized to a formula of descriptively outlining regional stratigraphy.
There was no mention of theory, of Scripture, or of cosmogony.[9]
These volumes of the *Transactions* were influentially reviewed
(e.g. in the *Edinburgh review*) by the Society's own members, who
constructed a mythic history of the science, pronouncing that
before the Society's founding all had been speculative darkness,
whereas, since then, the science had undergone Baconian reformation ([Playfair], 1811; [Fitton], 1817a & b).

Whether the Geological Society might have been more socially
open than it was, its success on its own terms is clear. When
Charles Babbage in the 1830s bemoaned the decline of science in
England, he singled out the Geological Society – and geology
in general – as exceptions (1830: 45). Membership flourished.
Acquiring adequate premises and building up a library and collection gave the Society drawing power (H. B. Woodward, 1907:
73f.). Its debates were lively; its *Transactions* definitive. For the
next couple of generations, no important English geologist of
social standing declined to become a member. English geological
debate was centrally conducted within the Society's ambit.

The Wernerian Natural History Society in Edinburgh (f. 1808)
provides important comparison and contrast.[10] It too seceded as
a specialist society from a general parent – the Royal Society of
Edinburgh. Like the Geological Society of London, it too proved
a highly cohesive group. But whereas the closeness of the London
society was ensured by a shared commitment to co-operative,

empirical, untheoretical science, the Wernerian Society's unity lay in loyalty to the geognostical system of Werner, under the direction of his pupil, Robert Jameson. Many members of the Wernerian Society were Jameson's pupils, or ex-pupils, and it flourished and decayed in pulse with his own interests and Wernerian geognosy. It was more like Jameson's research seminar than a society of equals. The 'Queries' circulated by the Society sought corroboration of the Wernerian system (*Phil. mag.*, 1808: 282). The society made a successful appeal to many Scotsmen who did not have time, money or leisure for such thorough involvement in geological inquiry as the leading members of the Geological Society of London, but who could carry out well-directed, small-scale research tasks. With the demise of Wernerian geognosy, however, the society lost its rationale, and it decayed in the 1830s.

The founding of the Royal Society in 1660 had provided a continuing base for the general investigation of the Earth. With the breaking away of the Geological Society of London in 1807, English geology achieved independence and the institutional basis for its future development.

## The Scottish School of Geology

I have argued in this chapter that late eighteenth-century English society provided the social basis, not just for a great expansion in Earth science, but for its channelling into the intellectual and institutional forms of geology. But late eighteenth and early nineteenth-century Scotland also supported a cluster of extremely eminent Earth scientists – John Walker, Robert Jameson, James Hutton, John Playfair and Sir James Hall – and contemporary Edinburgh, as the home of Hutton, has been claimed the cradle of modern geology. I wish now to analyse the social basis of this great late eighteenth-century flowering of Scottish Earth science, both in respect of the development of science in Scotland, and by comparison and contrast with the English experience. In doing so I hope to illuminate major distinctions between English and Scottish Earth science. For, if we were to say that the science of the Geological Society of London became the *locus classicus* of contemporary geology, then geology in that sense was little practised in Scotland. Hutton himself did not often use the term. Jameson rejected it in favour of Wernerian 'geognosy'. In some ways, indeed, Scotland at the turn of the century witnessed the last stand of the old traditions of cosmogonical theory and of the natural history of the Earth.

In a justly famous essay, Sir Archibald Geikie discussed 'the Scottish school of geology' (1871). This he identified with James Hutton and his friends, claiming that this Scottish school was in effect the foundation of modern geology in the British Isles. Without at all questioning the profound originality and importance of Hutton, I wish from a very different perspective to suggest, however, that Huttonianism was *not* the characteristically *Scottish* school of geology, and furthermore that Hutton's *Theory of the earth* only rather marginally arose from the traditions out of which British geology grew, and was scarcely acknowledged as geology by the leading geologists of early nineteenth-century England. To grasp what *is* the Scottish school of geology we must briefly examine Earth science in Scotland in its wider parameters of time and community.

Before the Act of Union (1707), under stimuli such as the mechanical philosophy, Baconianism and the founding of the Royal Society of London, Scotsmen like George Sinclair, Andrew Balfour, Robert Sibbald and Robert Wodrow began to cultivate an interest in Scottish mineral and natural history, leaning towards virtuoso observation and collection, often with utilitarian motives. This development, however, proved even less self-sustaining than in England, for lack of regular teaching in the universities and scientific societies. Interest in the Earth was subordinate to professional preoccupations in medicine and the Church. Wodrow abandoned his fossil collecting as he rose in the ministry. The collections of Balfour and Sibbald were dispersed after their deaths. In an age of marked growth of local investigation and theories of the Earth in England, the Scots who made greatest contributions were chiefly those who had taken the high road south – e.g. Sir Robert Moray (Robertson, 1922), or John Arbuthnot.

During the next seventy years investigation of the Earth found institutional, though not yet professional, recognition, and expanded accordingly. The many newly founded improvement societies put agricultural advancement, mineral discovery and exploitation, quarrying and coal-mining high on the agenda (McElroy, 1969). The Philosophical Society offered free mineral analysis in the 1740s (Shapin, 1974a: 9). John Walker toured the Highlands and Islands in the 1760s under the patronage of Lord Kames, reporting on mineral, metal and building stone resources (Walker, 1764; Scott, 1966: xix). Foremost patrons of Earth science included improving land and mine-owning aristocrats and gentry like the Earl of Hopetoun, the Duke of Queensbury, Sir John Clerk of Penycuik and Lord Dundonald (Cochrane, 1793; Sir A. Murray, 1732; Clerk, 1892). This utilitarian impulse gave Scottish investigation a secure footing for the first time, and established a long-lasting alliance between land-owning gentry, scientific societies and Earth science (Shapin, 1974b). But its high price was a period almost devoid of *ideas*. Nevertheless, the growth of chemistry, mineralogy and natural history lecturing in the universities – with Cullen and Black in Edinburgh, and Francis Shene at Marischal College, Aberdeen – began to consolidate a general systematic study of the Earth (Clow and Clow, 1952; Ritchie, 1952).

The 1780s saw two crucial developments. A continuous tradition of high quality lecturing was inaugurated at Edinburgh, starting with John Walker, followed by Robert Jameson. And the Royal Society of Edinburgh was founded. These ushered in Geikie's great age of Scottish geology, by producing institutions – and counter-institutions, like Jameson's Wernerian Natural History Society – which focused intellectual debate, provided educational opportunities and caught the public ear. Thus the trajectory of the development of Earth science in Scotland moves across several generations, within and parallel to the unfolding of the larger destiny of the Scottish Enlightenment (cf. Phillipson, 1973). We do not understand Scottish Earth science by focusing on Hutton alone, nor understand Hutton except in context of his wider community.

This is particularly important, given that even in their own generation, Hutton and the Huttonians were a clear minority, who had no monopoly of talent or followers. True, Huttonian geology had a powerful grip on the Royal Society of Edinburgh. But there never were many Huttonians, and even Hutton's leading defenders, Playfair and Hall, dissented strongly from him in important issues. Jameson had far more followers, who, in the Wernerian Natural History Society, together carried out a research programme of descriptive geognosy and mapping of real importance for the development of Scottish geology. Beyond his immediate circle, Hutton won few converts. The members of the Edinburgh Academy of Physics leant a little more towards Werner, though the early geological notices of the *Edinburgh review* pronounced a plague on both their houses.[11] John Murray, in his anonymous *Comparative view* (1802), and James Rennie, in his unpublished 'Comparative view of the Huttonian and Wernerian theories of the earth' (1816), both espoused a qualified Wernerian position. Furthermore, Hutton's work was ambivalently regarded and little read by nineteenth-century Englishmen and Continentals, even by actualists, till Geikie's championing. Hutton was not typical of Scottish Earth science.

What then *was* characteristic and distinctive about Scottish Earth science? Firstly, as contrasted to England and France, it was socially homogeneous. Scottish geology was Edinburgh geology. Edinburgh geologists eagerly banded into scientific societies and subscribed to schools of geological thought (cf. McElroy, 1969). Their geological allegiances were bandied around as part of broader commitments to common philosophical, religious and scientific paradigms. The general trend of English Earth science had been towards disaggregation, localism and intellectual independence. The Scottish milieu – of university education, Edinburgh societies, and a small, broadly ranging, well defined intelligentsia – kept Scottish Earth science highly 'philosophical' (Davie, 1961).[12] And this philosophical stance bred the continuing and impassioned contentiousness and controversialism which was so obtrusive in Scottish geology at the turn of the century. 'I am a great friend to controversy and keep it up here as much as possible', boasted Sir James Hall (1812). The rivalry between Huttonians and Wernerians in the Royal Society of Edinburgh was of a kind unknown to the Geological Society of London, spilling over into ill-natured personal wrangles, for example over the use of the University Museum. Thomas Thomson's *Annals of philosophy* was the theatre of prickly, provocative reviewing, and full-blooded controversy was encouraged by the *Edinburgh review* (Cockburn, 1856: 204–6; Chitnis, 1970; Clive, 1957: 30f.). Fierce Scottish geological controversialism probably in the long run had an inhibiting effect upon the science. Premature polarizations locked men in sterile postures, whereas English evasion of deep theoretical issues provided a basis for co-operative normal science to flourish.

In two other major respects the Scottish social and intellectual milieu shaped Earth science in directions significantly different from the science in England. The first was an overriding, deep and continuing concern for utility. This was a response to the all-pervasive need to modernize the backward Scottish economy. It took various forms. Joseph Black and Thomas Hope amongst other leading scientists undertook mineral analysis (McKie and Robinson, 1970; Clow and Clow, 1952: chs xvi–xvii). John Williams's *The natural history of the mineral kingdom* (1789) was the most substantial British treatise of its age on coal finding and mining. Williams sought Lord Buchan's patronage for developing Scottish mineral industries,[13] just as John Walker (1808) and Robert Jameson (1800) travelled the Highlands and Islands with a view to stimulating quarrying and mining, and Rudolf Erich Raspe had covered the same ground in behalf of Sir John Sinclair (Carswell, 1950: 12–31; Mitchison, 1962: 12f.). 'My leading idea in natural history is to render it subservient to the purposes of life', wrote John Walker (1789). At no point were leading English naturalists so genuinely concerned with the economic aspects of Earth science as their Scottish counterparts.

Hutton, however, is the Scottish exception. Of course, he made comfortable profits out of sal ammoniac manufacture, took an interest in steam engines and canals, and wrote a treatise on agriculture, whose expressed aim was to make 'Philosophers of Husbandmen and Husbandmen of Philosophers'. Yet the Huttonian theory of the Earth is totally unconcerned with utilitarian applications. Indeed, the very nature of his theory precludes practical use. Its idea of strata, for instance, had no predictive value for the discovery of coal seams and mineral sources. In fact Hutton could be loftily dismissive of utilitarian science:

Science, no doubt, promotes the arts of life; and it is natural for human wisdom to improve these arts. But, what are all the arts of life, or all the enjoyments of the mere animal nature, compared with the art of human happiness – an art which is only to be obtained by education and which is only brought to perfection by philosophy. (1794a: v)

The Royal Society of Edinburgh, Huttonian in its geology, did not publish practical geology, unlike the Wernerian Natural History Society to which Robert Bald, Scotland's leading coal engineer, was a prominent contributor. In this respect, Hutton was swimming against the tide of Scottish geology.

The second defining feature of Scottish Earth science was an *enduring* concern, even into the nineteenth century, to realize it as 'the natural history of the mineral kingdom' (to borrow the title of Williams's book) within a 'philosophy of natural history' (the phrase is William Smellie's, 1790). This comprehensive natural history approach was common to Balfour, Sibbald, Shene, Cullen and Black, but then still to Walker, Hope, Jameson and the Wernerian Natural History Society – a significant name. The Scottish tradition was a highly systematic version of the seventeenth-century English natural history approach to the Earth, unique for its resilience throughout the eighteenth and nineteenth centuries. Earth science was still taught at Edinburgh University in the 1860s by a professor of natural history.

The tradition aimed to articulate investigation of the mineral kingdom within a comprehensive vision of the three realms of nature – as is particularly evident in the life-work of Walker and of the Wernerian Natural History Society. Such a synoptic ambition for natural history presupposed a total philosophy of the created order of Nature, hingeing on a metaphysics and natural theology of design, the Chain of Being, teleology, order, hierarchy and mind–body dualism. Walker wrote a memorandum in the 1790s entitled 'Occasional Remarks' in which, as a Common Sense philosopher, he defended such a traditional Creationism against the philosophical and scientific errors of materialism, monism, naturalism, reductionism, necessity, Deism, Humean views on causality, and belief in the eternity of matter and the Earth.

This natural history of the Earth – whether in laboratory or in the field – focused on identification of rocks, naming and classification. Walker pointedly saw himself as heir to Linnaeus, and Jameson was a pupil of Werner (Scott, 1966: 253f.). As Walker's and Jameson's questionnaires show, the tradition was committed to a mentality of collecting, fact gathering and arrangement (Sweet, 1972; Scott, 1966: 219–20). Underpinning all, was a Baconian epistemology and methodology, explicitly hostile to naturalistic speculation. As Walker warned in his lectures upon the strata in the 1780s, 'I would not wish to be thought to deliver anything like a Theory, but merely a natural history of the earth' (Scott, 1966: 180). Late seventeenth-century English Earth science was also deeply rooted in these traditions, but English investigation had shifted away towards a local geology based on stratigraphy and lithology. But this happened much less so, and more slowly, in Scotland.

The reason lies in the needs, opportunities and constraints of Scottish university science teaching. Almost all great Scottish Earth scientists were teaching professors (Morrell, 1971a; 1974). A simple, orderly, descriptive account of the mineral kingdom set out within the natural history method was appropriate to the course demands of Scottish students. Such a need was hardly felt in England, however, where geology did not significantly develop out of university teaching situations, but rather from traditions of local amateur observations – the parson naturalist and the traveller naturalist – and where, therefore, the systematic natural approach became redundant. But once again, Hutton is the exception to the Scottish norm – and to the English. Hutton took Earth science out of natural history and thrust it firmly back within the physical and chemical sciences and the theory of matter. He saw it as akin to chemistry, a science of laws, powers, process and equilibrium. Hutton rejected the static hierarchy of the Great Chain of Being and substituted a dynamic, interactional view of Nature. He abandoned Baconian empiricism and Scottish Common Sense philosophy as epistemologically simple-minded, and developed his own philosophy of Mind, disparaging the credulous dilettante and arguing that the book of Nature was not, indeed, open to all to read.[14]

I have argued that Scottish Earth science differed markedly from that developed in England; and that it reflected the situation of a closed community of Edinburgh naturalists, gearing their science to university teaching needs, within an economy which urgently demanded technological improvements. Hutton produced a theory of the Earth which was dramatically *untypical* of

Scottish Earth science. He was the exceptional geologist, precisely because he stood *outside* the usual structures of Scottish geology. He spent formative years in Enlightenment Paris and Leiden. He never held a university post. He never lectured – hence perhaps his infamous style. He was never employed to prospect for minerals or to survey Scotland. He was not famed as a collector. He did not publish early in life or have literary ambitions. If the Scottish intellect was 'democratic' (Davie, 1961), Hutton was an aristocrat of the mind. Add to this Hutton's own pronounced personal eccentricities, hinted at in Playfair's biographical account (1805), and his originality is unsurprising. Hutton's originality was that of one who worked *outside* the Scottish tradition of Earth science; but *within* certain Scottish traditions of philosophy, as I shall explain more fully when I discuss the context of Hutton's geo-philosophy in chapter 8.

The social milieu determined the different traditions of Earth science which flourished in Scotland. The traditional forms of Earth science were more persistent in Scotland. Scotland did not produce a full-scale physical theory of the Earth until such theories were obsolescent in England. Yet, because of the synoptic ambitions of the Scottish Enlightenment, Hutton's *Theory* was a true product of many of Scotland's intellectual currents. Meantime, investigation of the Earth through natural history remained relevant longer in Scotland than in England on account of the strength of Scottish university teaching and the weakness of Scottish provincial science.

# 7

# A reformation of knowledge

## Geology: A new focus on the Earth

In this chapter I will examine how particular areas of the scientific investigation of the Earth were profoundly transformed in Hanoverian England. But I will also investigate how the *ensemble* of investigations was abandoning the natural history framework of expectations, which realized particular objects against a universal grid of Nature, and was developing a dynamic, relational, understanding of objects and processes within a view of the integrated development of the Earth; was becoming in fact *geology*. The formation of a new kind of science was at the forefront of contemporary consciousness:

Geology is principally distinguished from *Natural History*, which confines itself to the description and classification of the phenomena presented by our globe in the three kingdoms of Nature, insomuch as its office is to connect those phenomena with their causes. (de Luc, 1793–4: 231–2)

Aspects of the natural history of the Earth were being rejected as obsolete for the new geological vision. William Smith argued that preoccupation with collecting and classifying fossils should yield to understanding them within the geological order of the strata – which, to his mind, gave fossils real significance for the first time (Cox, 1942–5: 12). Smith himself was then 'trumped' by the even broader geological view of Cuvierian palaeontology, which saw fossils not merely in relation to strata, but within the perspective of the entire history of life.

In a parallel way, mineralogy could become comprehended within geology, as the static within the dynamic, the building bricks within the fabric. In an image which was to become commonplace, Richard Kirwan called mineralogy the letters of

Nature, which, written up into words, became geology (1799: Preface). Significantly, during his own working life, Kirwan seems to have progressed from *Elements of mineralogy* (1784; 1794–6) towards *Geological essays* (1799). Geologists began to castigate mineralogists for narrowness and artificiality. A few years later William Whewell graphically summed up the relationship:

The connection of mineralogy with geology, is somewhat of the nature of that of the nurse with the healthy child, born to rank and fortune. The foster-mother, without ever being connected by any close natural relationship with her charge, supplies it nutriment in its earliest years, and supports it in its first infantine steps; but is destined, it may be, to be afterwards left in comparative obscurity by the growth and progress of her vigorous nursling. (Weld, 1848, ii: 245)

It became commonplace to explain geology's late emergence as due to the fact that other, auxiliary sciences had needed to develop first as under-labourers to geology (Playfair, 1802: 5; Bakewell, 1813: ix).

The desire and the ability to organize thought and fieldwork within a geological vision was the culmination of trends in observation and philosophy of Nature. New observation continually revealed features which old outlooks and theories could not satisfactorily explain. Seeing the Earth as the order imposed at the Creation and modified by the Deluge could no longer convince those familiar with the complexities uncovered by fieldwork. As Robert Townson remarked, fieldwork revealed multiple changes in the state of the Earth, proving 'that our globe, or rather its surface, is not the simultaneous formation of the Omnipotent *fiat* but the work of successive formation and subsequent changes', which are 'strong hints, or rather indubitable proofs, of great revolutions' (1798: 4). Richard Sulivan acknowledged the same:

Thus succeeds revolution to revolution. When the masses of shells were heaped upon the Alps, then in the bosom of the ocean, there must have been portions of the earth, unquestionably dry and inhabited; vegetable and animal remains prove it; no stratum hitherto discovered, with other strata upon it, but has been, at one time or other, the surface. The sea announces every where its different sojournments; and at least yields conviction that all strata were not formed at the same period. (1794, ii: 169–70)

In particular, strata complexities became ever more clearly

exposed, posing further problems. But further discovery of the order of the strata, and a growing capacity to trace their regularities over large tracts of land, also made it possible to devise satisfying accounts of the Earth, based upon the evidence of the strata, to account for these complexities. The strata became the focus of practical geological activity, and also the referent in relation to which other terrestrial phenomena were to be understood.

The meaning of specialist studies could be integrated within a geological view because by now a long time-scale was almost universally accepted. Sir William Hamilton's work on the successive eruptions and lava flows of Vesuvius had indicated that currently active processes within that volcano must date back at least some fifteen or twenty thousand years. Hamilton was proof against being labelled a radical, atheistic philosopher. Buffon's quantified estimates of the age of the Earth provoked controversy rather than acceptance, but his *Histoire naturelle* (1749–) and *Époques de la nature* (1778) also ushered in an intellectual climate in which a long time-scale was in principle acceptable. William and John Hunter's work on the relation of fossil remains to strata explicitly spoke of long ages and great epochs in Earth history (W. Hunter, 1768: 36; J. Hunter, 1794: 409). Few might accept Hutton's or George Hoggart Toulmin's claim of an *indefinite* antiquity for the Earth, or share the enthusiasm of the *Analytical review* for an immense time scale when it approved of Cornelius de Pauw's demand that men should give

to this Globe its just right, an unbounded antiquity: admit that in the expanse of time, it may have undergone many very great changes as of ocean into continent, and of continent into ocean. (*Anal. rev.*, 1795: 336)

But the Neptunism of the pious de Luc and the Wernerianism of Jameson themselves required a long time-scale. In his popular *View of nature* Sir Richard Joseph Sulivan admitted that it could no longer be gainsaid that the Earth was of very great – though incalculable – antiquity (1794: ii, 189). The *Analytical review* truly remarked in 1797 that all recent theories required an extended antiquity for the Earth (238). Though suspicious about the conclusions which many would derive from acceptance of a long time-scale, and hostile to all attempts to put to it a figure, the Rev. James Douglas (1785) candidly admitted that fossil remains, the depth of strata – each of which must have taken ages to form – evidence of climatic change in cyclical fashion and so forth, all pointed unambiguously to an 'enormous' antiquity for the Earth. By the end of the eighteenth century, fieldwork and the new philosophy of Nature born of the Enlightenment fused to make *geological* interpretation of the Earth possible, desirable and necessary.

## Structure and surface

By the end of the eighteenth century almost all geologists had discarded the geomorphological theory of the complete natural stasis of the Earth. This was in part because the natural philosophical and theological imperatives which had created the 'denudation dilemma' had been disintegrating. But it was also because of facets of Nature spotlighted by observation. Confrontation with three particular natural features was undermining belief that past changes of the crust could not have been due to natural causation, and that actual physical causes were not competent to transform the Earth.

Firstly, investigation of strata – their depth, internal form, disruptions, and their fossils – was ever more clearly suggesting they must be the product of natural causes over time. It was now thought reprehensible theological obscurantism, the quietus of all philosophy, simply to resolve the problem of the strata by postulating a divine fiat at the Creation or the Deluge (cf. *Monthly review*, 1790: 541). For most, the strata gave undeniable proof of earlier changes in landforms. Sober interpretation of the evidence demanded acceptance that strata had been initially formed on the sea-bed, since they were largely composed of sedimentary detritus, and contained fossilized remains of shells and fish (Douglas, 1785). This entailed that land and sea had repeatedly changed place during Earth history, and supporting evidence was found when John Hunter among others pointed out the alternation of land and sea fossils embodied in the series of the strata (1794).

Furthermore, Neptunists no less than Huttonians were agreed that strata bore marks of being composed of the detritus of former continents. Richard Kirwan, for instance, thought coal comprised the remains of primitive mountains of mineral carbon utterly eroded by the water and weather (1799: chs. v, vii). If all *Second-*

*ary* strata (as the Wernerians believed), or all rocks except granite (as Hutton believed), were composed of eroded materials of earlier rocks, the Earth must have suffered repeated extraordinary natural transformations of relief. Of course, one of the central bones of contention between Huttonians and Wernerians, Uniformitarians and Catastrophists, remained the question whether these processes continued to operate, either at all, or with the same intensity as in the past (Cannon, 1960a; Hooykaas, 1963; Rudwick, 1967). But both groups had transcended the fundamental denudation dilemma to the extent of accepting, when faced by the evidence of the strata, that at least in the past monumental *natural* physical forces had revolutionized the face of the Earth.

Secondly, close investigation of earthquakes and volcanoes similarly revealed fresh and forceful evidence of massive natural transformation of the crust. As already discussed, growth in scientific travelling by those embracing the Enlightenment philosophy of active Nature led many to grasp the integral place of these phenomena within the Earth's economy. Their empirical methods, and their meticulous observations, the extensive writings of such as Sir William Hamilton, enhanced by pictorial representations in both his *Philosophical transactions* articles and in the *Campi Phlegraei* (Rudwick, 1976), all helped to win credence amongst scientists and the educated public for the notion that volcanoes were not freaks of nature of which extravagant writers had offered exaggerated accounts, but positive forces in shaping the wider face of Nature. Especially important was Hamilton's emphasis that, though volcanoes and earthquakes, viewed superficially, were unnatural, terrifying destructive and irregular, yet regarded in a wide philosophical perspective they formed part of the regular and constructive operations of Nature. They were orderly and predictable, in that they operated in defined parts of the globe. They moulded the landscape by stages. Vesuvius and Etna had been gradually built up over many thousands of years by tens of successive eruptions. 'Nature, though varied, is certainly in general uniform in her operations', wrote Hamilton in an influential apothegm.[1] And lastly, volcanoes were a *constructive* agent of Nature: 'Volcanoes', Hamilton wrote, 'should be considered in a creative rather than a destructive light' (1786: 378). The effects of Vesuvius and Etna had been to enlarge the west coast of Italy and Sicily, not to throw the area into confusion. Other mountains might have been similarly formed (1769: 21).

The work of Hamilton and the many other investigators of volcanoes was of threefold importance. It focused attention on active forces of Nature, possibly the products of global central heat. Then, it showed that such forces continued to this day, which was important in underlining the continuity of the Earth's present with its past, and thereby encouraging actualistic methods. And thirdly, it encouraged massive and continued search for extinct volcanoes, which could be identified by the presence of lava flows, by similar lithological features such as basalt, or by landscape characteristics, such as conical hills with flattish tops (Hamilton, 1771: 6).

Between the 1760s and the 1780s, William Bowles, Sir William Hamilton, Sir John Strange, Augustus Hervey and Patrick Brydone were the Britons who pioneered investigation of Continental volcanoes. Sir James Hall later assumed the mantle. Hall travelled extensively through the extinct volcanic areas of Germany, noting formations of igneous rocks and volcanic landscapes, before proceeding to Vesuvius and Etna. He had already imbibed Hutton's views on the Earth's activity before making these travels, but his work in particular on the lava flows of Etna convinced him of the power of successive eruptions to create huge tracts of land, the slowness of the degradation of lava flows, and of the major role of volcanoes in the creation of the landscape (1783–4).

Such investigation spilled over into the revolutionary conception that the British Isles harboured ancient volcanoes, identifiable through landscape features and/or formations of lava, pumice, tufa and basalt. The Giant's Causeway obviously attracted attention. But the prismatic basalt formations on Staffa and other Scottish Isles now began to be connected with volcanic action, or at least with flows of molten rocks, whether on the land surface or the sea-bed (Pennant, 1771). Thomas West thought he had discovered volcanoes near Inverness (1777). Abraham Mills found them all across Scotland (1790), as did the French geological traveller, Faujas de St Fond (Geikie, 1907, ii).[2] John Whitehurst had believed Derbyshire toadstone to be evidence of the universal volcanic eruption. When William Maton discovered amygdaloidal rock *in situ* near Exeter on his journey, he felt, as a Neptunist, the need to spell out in some detail why he exceptionally did *not* associate such rocks with past volcanic activity (1797, i: 95–7).

Additionally, intrusions and extrusions such as whin dykes and sills in northern England and Scotland were increasingly being investigated as evidence, if not of volcanoes, at least of active igneous forces which had transformed the physical appearance of the Earth's crust. Furthermore, one school of mineralogists came to interpret the crystalline or massy nature of many rocks as indicating an igneous, if not necessarily volcanic, origin. Hence, the more rugged regions of the British Isles were increasingly seen as having been thrust up from the bowels of the Earth. For all these investigators, the philosophy which saw the globe as passive and changeless was dead.

The third natural feature to attract growing awareness was the regular denuding force of sea, weather and rivers, and of more catastrophic, but nevertheless natural, débâcles. Admittedly, Hutton needed the backing of an entire philosophy of Nature to argue a case for profound *fluvial* erosion. But to the growing band of travellers across Alpine Europe and the Americas, the capacity of torrents, glaciers and climate to destroy mountains and to gouge out valleys became patent and beyond dispute (cf. Sir J. Hall, 1783–4, July–Aug.). Wernerians, like Jameson, Hailstone and MacCulloch, no less than Huttonians could accept present landscape transformations by such natural causes (Davies, 1969: 123–4, 194, 226, 235f.).

By the late eighteenth century, then, most geologists were convinced that widespread denudation was actually occurring. But the condition of the Earth as newly understood required that massive erosion and redistribution of materials throughout the globe had occurred in the *past*, to explain the very contouring of the crust, and the denudation of former continents from whose detritus present Secondary strata had been formed. One Deluge – even if a unique event of this kind were still scientifically credible – could not explain the multiple, diverse and complex phenomena. To interpret landforms, it had become necessary to invoke a succession of natural débâcles, and in addition the continuous operation of climate and oceans. When Sir James Hall spelt out his theory of débâcles to account for landforms, Leonard Horner evidently thought that he was reneguing on his Huttonian past, and reverting to a theory of a single universal Deluge. But Hall explained this was not the case at all:

> I shall be most happy to show the Diluvian facts to Mr. Horner. I must have expressed myself very ill in some part of the paper to give rise to the opinion which he ascribes to me, that only *one* such cause has acted. On the contrary, it is my firm belief thousands of such actions have taken place in succession; nay, I conceive them to have arisen as the necessary consequence of all the Huttonian elevations. (Sir J. Hall, 1813)

In other words, just as the effects of volcanic eruption and earthquakes on landforms could be conceptualized as more natural through a perspective which accepted them as repeated and gradual, so also for denudation forces. The absolute polarity between miraculous catastrophes outside the course of Nature and a purely stable Earth could be broken down by introducing the concept of mighty constructive and destructive events, seen nevertheless as orderly, local and within the course of Nature (cf. Rudwick, 1971).

So, ever more evidence for apparently natural, necessary and regular landscape changes came to light, which could be readily incorporated within their philosophies of the Earth by naturalists of many theoretical persuasions. Certain thinkers, such as James Hutton and Erasmus Darwin, saw activity, process and change as essential to a wisely designed Earth economy: 'this active scene of life, death and circulation' (Hutton, 1795, i: 281). Only thereby could destruction be repaired, and the Earth's habitability and progressive nature be maintained. Others allowed that more limited change was essential to the overall conservation and benefit of the Earth (Davies, 1969: 119–20). Alternatively, Wernerians thought that the planet's earlier history had been a theatre of turbulent physical forces and enormous landform change, but that as it developed, it had become progressively more quiescent and stable. Much this kind of view had been taken by Whitehurst. Active forces were to be eagerly embraced in the nineteenth century by Christian writers seeking evidence of the discontinuities of life and Nature. Within nineteenth-century structures of viewing the Earth, both Uniformitarians and Catastrophists presupposed enormous change across the face of the globe.[3]

One wing of philosophers of the Earth, however, clung to the vision of the terrestrial economy as absolutely changeless – all those who thought any concession to naturalistic Enlightenment views of an active Earth was the thin end of the atheistic wedge. These included 'scriptural-geologists' like Philip Howard (1797). Some natural historians, like John Walker, educated in a static vision of Nature, were deeply hostile to the Enlightenment underpinnings of volcanism and to all possibility of active terrestrial forces.[4] And faced by Hutton's *Theory* and the French Revolution, de Luc and especially Kirwan emphatically set their faces against an Earth in permanent revolution. De Luc in fact had stressed the present quiescence of the Earth from his earliest geological writings, which were even then a polemic against the *philosophes* (1779). In fact even he *had* earlier been willing to attribute quite an active role to volcanoes and denudation while maintaining that the present continents were recent (de Luc, 1779, iii–iv). But by the 1790s he was not only more emphatically denying current denudation, but overtly associated such beliefs with an atheistical cosmology. De Luc's responses were those of an old man trapped within his early perceptions that Christianity hinged on a stable, designed habitat (1790–1).

Significantly, de Luc's response found no followers amongst the next generation of eminent investigators. Subsequently, Christian opponents of Hutton could accept his chapters upon denudation as tolerably valid (Greenough, 1819a; Rudwick, 1963). By the early nineteenth century there was no one left who was still prepared to contest that the Earth had undergone considerable changes. Rather, debate centred on the *order* of the causation and the *pattern* of the activity (Page, 1963; 1969). Acceptance of complex active processes within the economy of the Earth – past and present – broke down the seventeenth-century structure of static natural history punctuated by unique and miraculous revolutions. By the close of the eighteenth century landforms were being investigated within a geological world view.

### Fossils and life

Landforms and strata could be successfully studied by gifted amateurs with energy, wealth and method. Britain produced competent geomorphologists and stratigraphers. The problems of fossils and minerals, however, were more technical, and required a combination of professional training, institutional co-operation, apparatus and equipment rarely found in Britain (cf. Parkinson, 1804–11, iii: Preface). Britain's palaeontologists and mineralogists were markedly inferior to those in France and Germany, from whom their work was largely derivative.

The early eighteenth century had moved towards consensus that figured stones were genuine organic remains, probably the relics of the Deluge. A natural history tradition of identification and classification had been established. Through the century such work continued. As study of strata became more precise, it became difficult to accept that all fossils could owe their origin to the same time and cause, the Deluge. Whitehurst could imply that life, like the Earth, had a high antiquity (1778). Organic remains had been caught up in sediments on the sea-bed at distant periods. Subsequent revolutions of the globe had transformed sea-bed into solid continents.

During the last quarter of the century, this unspecific, but comprehensive, interpretation of the history of life through fossil evidence, first brought to light by reading the strata, now continued to develop. It reached its peak of coherence in Jean André de Luc. De Luc affirmed that the deepest strata were unfossiliferous, indicating that such rocks had been precipitated before the creation of life (1831: 56f.). He recognized a broad range of subsequent sedimentary strata containing fossil shells, plants and fish (118f.). Vertebrate bones in various degrees of petrifaction were to be found amongst more superficial deposits, but scarcely any, if any at all, fossilized human remains (20–2). For de Luc, this was sound evidence for the general successive Creation of life as set out in the Bible (47f.).

Neither de Luc, nor any of his English contemporaries, explored the relationship between fossils and the history of the Earth and of life with greater discrimination. Erasmus Darwin realized similar data in a physical evolutionary framework (1803: cantos ii–iii). Scots geologists had less opportunity for fossil study, and Hutton took little interest in it (Eyles and Eyles, 1951: 320). Most English mineralogists, with their chemical background, were not particularly concerned about fossils. Potentially important discoveries, such as Joshua Platt's finds of fossil vertebrates in the

Stonesfield slates (1758), and the highly proficient studies of John and William Hunter on petrified mammalian remains (1768; 1770; 1794; F. W. Jones, 1953) made little impact for lack of a community with the requisite blend of geological and comparative anatomical competences.

Fossil study advanced in England, not through using fossils to explain the history of life, but within an essentially different context. The milieu was not, as in France, the metropolitan scientific elite, but the network of provincial professionals, whose vision of fossils was undoubtedly narrow, but whose acquaintance with them was great, and who had a sharp eye for grasping how fossils could solve their immediate problems.

Typical was William Martin, who spent much of his life collecting and describing local fossils, of which he published accounts in his *Figures and descriptions of petrifications collected in Derbyshire* (1793) and subsequently in *Petrificata Derbiensia* (1809). His fossil studies left him dissatisfied with mere description within the natural history tradition. In his *Outlines of an attempt to establish a knowledge of extraneous fossils* (1809), he attempted to reform the direction of the science, by constituting fossils as an auxiliary to *geology* – though not as the key to the history of life: 'The study of *extraneous fossils* is confessedly useful to the *Geologist* – it enables him to distinguish the relative ages of the various strata, which compose the surface of our globe; and to explain, in some degree, the processes of nature, in the formation of the mineral world' (1809a: Preface, i). This statement epitomizes the meaning of fossils within contemporary British science. Liberated from natural history, fossil study became the handmaiden of stratigraphy, not the key to the integrated history of life and the Earth. James Parkinson concurred:

The study of fossil organized remains had hitherto been directed too exclusively to the consideration of the specimens themselves; and hence has been considered rather an appendix to botany and zoology than as (what it really is) a very important branch of geological inquiry...To derive any information of consequence from them on these subjects, [i.e. the formation and structure of the Earth] it is necessary that their examination should be connected with that of the several strata in which they are found. (1811: 324)

William Smith used fossils precisely within this framework (1796):

*Fossils* have been long studdied as great Curiosities collected with great
pains treasured up with great Care and at great Expence and shewn and
admired with as much pleasure as a Child's rattle, or his Hobby horse is
shown and admired by himself and his playfellows – because it is pretty.
And this has been done by *thousands who have never paid the least regard
to that* wonderful order and regularity with which Nature had disposed
of these singular productions and assigned to each Class its peculiar
Stratum. (Cox, 1942–5: 12)

Smith was not the first to recognize that strata could be identified
by organized fossils. Neither was his fossil knowledge very subtle,
for example of *why* certain fossils were regularly sited in particular
strata (Phillips, 1844: 28). Rachel Bush has recently argued that
Smith himself made scant use of his own principle of identifying
strata by organized fossils (1974: chs. ii–iv).

On the other hand, that was the trumpeted principle of his
published works, integrated into the very substance of his maps
and sections, even to the extent of having engravings of fossils
printed on paper coloured the same as the strata on the map. In
these printed works, Smith fully established the empirical correlation between strata and fossils. With Smith, as with Martin, study
of fossils was shifting from a natural historical to a geological
perspective, but was not yet palaeontological. Smith showed little
desire to reconstruct the animal and vegetable economies of which
the fossils were remains. He had very little interest in Cuvier's
comparative anatomical approach to fossils.

The state of British fossil science at the beginning of the nineteenth century was epitomized by the first volume of James Parkinson's *Organic remains of a former world* (1804). Parkinson
himself was vague about the age, relative or absolute, or the
original conditions of life of the fossils he was describing – talking
loosely of 'former worlds', 'antediluvian remains', etc. He complained he had

so little aid, speaking comparatively, from such works as have been published in this country, since although these subjects have engaged the
attention of several gentlemen, well qualified for their investigation; yet
the publications with which they have favoured the world, have chiefly
been detached essays, on some particular fossils only. (1804–11, ii: Preface,
v)

Hence he chose to write a popular work in epistolary form, to
appeal to a general audience. Fossils commanded a small portion

of the early volumes of the *Transactions* of the Geological Society of London, and a negligible place in the *Transactions* of the Edinburgh and Cornwall societies.

British fossil studies were, however, transformed by Cuvier and his French co-workers (Rudwick, 1972: ch. iii). Subsequent British naturalists acknowledged that it was to Cuvier alone – or perhaps to Blumenbach as well – that they owed their understanding of fossil vertebrate bones, and their role in the history of life, stage by stage.[5] Introducing the English translation of Cuvier's *Discours préliminaire*, Robert Jameson uttered what was to become commonplace praise of Cuvier: 'His great work on Fossil Organic Remains of which a new edition is now in progress is the most splendid contribution to Natural History furnished by any individual of this age' (1822: vii–viii). Jameson praised Cuvier's work, in conjunction with Werner's on the succession of the strata, for enabling the geologist to identify strata with more success. But further, he went on, it also showed the geologist

the commencement of the formation of organic beings, – it points out the gradual succession in the formation of animals from the almost primeval coral near the primitive strata, through all the wonderful variety of form and structure observed in shells, fishes, amphibious animals, and birds, to the perfect quadruped of the alluvial land; and it makes him acquainted with a geographical and physical distribution of organic beings in the strata of the globe very different from what is observed to hold in the present state of the organic world. (1822: vii–viii)

No native British worker had attained such a comprehensive vision. British publications on fossil reptiles and amphibia, begun by Sir Everard Home's plagiarisms from John Hunter (1817; 1818; 1819) and continued above all by Buckland (1823) and W. D. Conybeare and de la Beche (1821) were deeply within traditions pioneered by Cuvier. Admittedly, Cuvier's diluvialism earned him the enmity of Playfair (1814), and both George Bellas Greenough (1819a: 280f.) and Charles Lyell believed that his succesion of organic creatures was the unwitting slippery slope to Lamarckism. But all *practical* work on fossils, especially vertebrate remains, in early nineteenth-century England took its cue from Cuvier's vision and techniques.

This failure of any Englishman to be pre-eminent in the interpretation of vertebrate fossil remains from the time of, say, Peter Camper through to Cuvier encapsulates the limitations of the

British, amateur, decentralized naturalist tradition. With the possible exception of John Hunter, Britain boasted no single general
naturalist of the range and distinction of Buffon, Lamarck,
Blumenbach or von Humboldt. Men of that kind required a
public situation lacking in Britain. Hence English science did not
generate a broadly based, integrated, history of life, derived from
fossils. But English men of science were not behind Cuvier in
recognizing the meaning of fossils for understanding strata, and
this alliance was to become a mainstay of nineteenth-century
stratigraphy.

## Rocks and minerals

As with fossils, the fundamental systematic study of minerals and
crystals at this time came from Continental Europe, especially
from Sweden, France and Germany. In the mid-eighteenth century, most British publications had either been compilations or
translations popularizing Continental developments. This situation
continued through into the nineteenth. British investigators, however, were prominent in particular areas of the deployment of
mineral knowledge, especially in solving problems associated with
the strata and with landforms.

It is not surprising Britain was relatively laggard at formal,
systematical mineralogy. Mineralogy, one of the most complex of
the aggregative natural history sciences, is perhaps less immediately attractive to the amateur than botanical and zoological collecting. It requires patient and delicate technique, laboratory
equipment, large-scale collections. The main impulse to systematic mineralogy is pedagogic. In Scandinavia, Germany and
to some extent France, mineralogy was indeed taught as a major
part of state-organized training in the techniques of mines operation and management. Professorships were available for skilled
mineralogists. In *laissez-faire*, decentralized Britain, such posts
scarcely existed (Guntau, 1967; 1968; 1969; 1971; 1974). In
Edinburgh, John Walker, Thomas Charles Hope and Robert
Jameson taught some mineralogy, but within a general natural
history course rather than as a directly practical or professional
subject. Edward Daniel Clarke began teaching mineralogy in
Cambridge from 1809, but again scarcely for vocational purposes.
Laboratory facilities were minimal. Plans for establishing mines

schools and professorships in mineralogy were often projected, as also were schemes for setting up full-scale mineralogical laboratories (Porter, 1973b: 332–5). But they came to nothing. Through into the nineteenth century, in systematic mineralogy British publications continued to be digests of Continental treatises. Some were highly scholarly, such as Richard Kirwan's *Elements of mineralogy* (1784). The alterations and additions incorporated in the second edition (1794) testify how well-informed Kirwan was of Continental developments. Some were heuristic popularizations, such as the works of Robert Townson (1798) and Robert Bakewell (1819). Some, like Robert Jameson's, were almost direct translation of Wernerian works (1805; 1808).

In mid and late eighteenth-century Europe, immense strides were made by a succession of great mineralogists – Pott, Linnaeus, Wallerius, Cronstedt, Bergman, Werner, Leske. New techniques were developed – such as blow-pipe analysis, and the chemical tests devised by Cronstedt. Routine analytical tests for identification were standardized, the Scandinavian and French schools tending to pursue chemical analysis, Werner favouring the use of external characteristics as equally reliable but much simpler (Werner, 1774; Ospovat, 1969). But of supreme importance were attempts to establish comprehensive classification systems. This was a matter of urgency, with the continual discovery of new minerals, and the painful awareness that the mineral kingdom – unlike the vegetable and animal – lacked sharp natural boundaries (Hooykaas, 1950; Ospovat, 1969). Linnaeus attempted a typology built upon the same framework as his classifications of the other two kingdoms, which John Walker popularized in Britain, but it was discarded as too artificial (Scott, 1966: 223f.). Rival systems sought more practical distinctions, built upon the maximum number of heuristic tests (Oldroyd, 1974b; 1975; Oldroyd and Albery, forthcoming).

But the development in the analysis and classification of the mineral kingdom which had greatest bearing upon Earth science was the move to base a natural classification of rocks on their location within the crust. Such a classification had the appeal of being 'natural' and 'real', and could also be a practical tool for finding minerals and for identifying those found *in situ*. Lithological classifications of this kind were developed in particular by

Torbern Bergman, in his *Physisk Beskrifning öfver Jordlotet* (1766; Hedberg, 1969), and by Werner in his *Kurze Klassifikation* (1787). These influential works crucially sought to understand the distinctions between particular rocks and minerals in terms of their different situation in the order of the strata. Further-more, they explained *why* differently situated rocks had distinct chemical and mineral compositions, through a sequential physical and chemical theory of the successive deposition of the strata. Bergman and Werner focused the growing perception that the lowest formations tended to be crystalline or massy, non-fossilifer-ous, often ranged at considerable angles to the horizon, and apparently chemically deposited. The rocks superincumbent on them were less crystalline, and had been mechanically deposited out of sediments – probably of primary rocks – and tended to be fossiliferous. Mineral expertise was being used to give a theoretical dimension to the empirically known order of strata.

This outlook within the mineralogical and lithological analysis of rock-forms was immensely important for the emergence of geology. Lithology might have remained a laboratory science of substance analysis based on natural history. Instead, under the guidance of Bergman and Werner, and then as popularized in Britain by Kirwan, Townson, Jameson, William Phillips, Bake-well, John MacCulloch etc., its attention became directed onto the mineral history of *this particular globe*. Future geologists might vociferously complain that Wernerian mineralogists assumed that the *only* proper tool for comprehending the history of the strata was mineral analysis.[6] But all had now been sensitized to the fact that dilemmas of Earth history – such as the aqueous or igneous origin of rocks, the intrusive or deposition theory of veins – could be explored through meticulous lithological analysis.

This was especially true as late eighteenth-century mineralogy provided a sequential, developmental model of Earth history, dividing the rocks into fundamentally different tangible characters according to the epoch at which they were laid down, compre-hended within some kind of explanatory mechanism, such as the diminishing waters of a universal ocean. As a descriptive theory, it had considerable heuristic value for geologists. It underpinned confidence in the order of the strata over wide stretches of terrain, and hence in the ambition of tracing them (Jameson, 1822: Preface).

Furthermore, Wernerian mineralogy offered exact criteria for identifying rocks and minerals, sufficiently simple to be used in the field. Identification of English Secondary rocks was relatively easy, and could be accomplished to perfection by a man of no mineralogical training, such as William Smith. But the highly confused areas of Devon and Cornwall, Wales, Scotland and much of Ireland were altogether different, as Smith's own patchy understanding of these locations suggests. Here the meticulous techniques of analysis developed by mineralogy were invaluable. Mineralogists, particularly those trained in Wernerian methods such as Robert Jameson, Berger, William Phillips, John Mac-Culloch, Charles Hatchett and William Maton, were naturally attracted to such areas and were most adept at unravelling their highly complex lithology.[7] Mineral lithology was furthermore putting the straight backbone of a strict terminology into the amorphous body of vague, local and traditional rock names. Wernerian nomenclature took firm hold. Many British geologists, such as Buckland, Conybeare and Greenough, who all rejected core Wernerian dogmas, nevertheless welcomed his mineral and geological terminology. So marked was this conversion that the adoption of the 'harsh Germanic geognostic vocabulary' was one of the *casus belli* of the war between Farey and the Geological Society of London (Porter, 1973b: 325–6).

Most importantly, as the intellectual framework of late eighteenth-century lithology became projected onto an interpretation of the rocks, it undoubtedly encouraged mineralogists out into the field. The mineralogists of mid-eighteenth-century England – Forster, Hill, Edwards, da Costa – were not fieldworkers. With their improved laboratory and chemical techniques, their successors might also have stayed in their closets (Townson, 1798: 114). But in fact, they became a much travelled fieldworking company, with Arthur Aikin active in Wales, William Phillips, Maton and Berger in Cornwall, Hatchett and MacCulloch touring over the country. For the more traditional, fieldwork chiefly yielded samples for practising fundamentally classificatory techniques. Thus William Maton, investigating strata near Exeter, dissociated himself from the '*Vulcanist*' geologist who would pronounce on their origins, and concluded that 'the cautious mineralogist' like himself 'will set it down among amygdaloidal earths without venturing to speculate on its origins' (1797, i: 96). This concern to identify rather than to examine history typifies the mineralogist's initial style of fieldwork. But some investigators, as I shall show in the next section, became increasingly discontented with the limitation of a purely mineralogical approach.

Especially because of the Lavoisierian revolution in chemistry, mineralogy became the battleground of debates on the theory of the Earth. Diluvialist and Neptunist theories – for which fossil and landscape evidence had been paramount – now also consolidated themselves on claims of the composition – hence origin – of strata, veins and dykes. Hutton maintained against the 'universal ocean' theory that siliceous rocks, being neither soluble in water nor porous, could never have been deposited from aqueous solution (1795: ch. vii). Kirwan and Jameson replied that because pure quartz was not fusible by fire, granite could not have had an igneous origin (Kirwan, 1791; 1802a; Sweet, 1967).

In particular, controversy over the role of volcanoes in the Earth's economy became translated into the mineralogical analysis of basalt, lava, tufa, pumice and related rocks. Volcanists claimed that basalt outcrops were mineralogically continuous (as well as geomorphologically similar) to rocks which were undeniably part of lava flows or associated with volcanoes (Strange, 1775a; 1775b; Raspe, 1771). At first this rested upon mere observations and comparison. But soon techniques were devised for mutually converting basaltic and vitreous substances, through controlled experiments in heating and cooling. Analogy with industrial processes was important to the volcanist case. Experiments showed that fluid glass, if cooled very slowly, became crystalline. This – it was argued – provided good experimental confirmation of why dykes of basalt generally exhibited crystalline structure, whereas other basalts, inferentially extruded as lava-flows, were sometimes vitreous.[8]

Of course, English Neptunist mineralogists devised their own counter criteria and experiments to demonstrate subtle chemical differences which obtained between basaltic rocks genuinely associated with volcanic activity and quite distinct dykes and sills of trap (i.e. basalt of aqueous origin) (Kirwan, 1799: 258f.). One of the most powerful arguments of the Wernerian mineralogists was that basalt could be found physically, locally and chemically graduating into other rocks undeniably of aqueous origin, such as porphyry, serpentine, wacke and granite.

This argument, however, came to be turned against its own authors by the growing group of British 'Plutonists' – Hutton, Hall, Beddoes, Gregory Watt – who saw all crystalline and Primary rocks as of igneous origins. Arguing that basalt had been proven igneous, they affirmed that the local and chemical affinity of basalt with other crystalline rocks including granite was good *prima facie* evidence for the igneous origin of granite and other crystalline rocks (Beddoes, 1791). Though Hutton himself shunned experimental chemical proof of his theory of igneous origins, preferring analogical arguments derived from local position and appearance (cf. Sir J. Hall, 1805: 45), the younger generation of Huttonians, notably Thomas Beddoes and Sir James Hall, had no such qualms about experiments. Hall in particular persuasively demonstrated in a distinguished series of experiments the interconvertibility of glassy and basaltine materials, and the fusibility of granite.[9] He also provided experimental confirmation of Hutton's contention that limestone and marble could have been fused by heat, without calcination, under extreme pressure. Jameson, amongst others, replied from the Neptunist point of view, reinterpreting the meaning and the relevance of these experiments (1795–6). Other mineralogists, such as John Murray, sought to arbitrate on basalt and granite and the derivative problem of the origin of veins ([Murray], 1802; cf. *Edinburgh review*, 1803: 391–8). Between about 1790 and 1810, the raging controversy between Huttonians and Wernerians was conducted in terms of alternative mineralogies.

Thus mineralogy contributed to geology a developmental Earth history, closely linked to the lithological understanding of successive formations, and also precise techniques for field identification, particularly of Primary formations, metamorphosed rocks, veins, and dykes. From the early nineteenth century onwards, however, mineralogy and geology became more clearly demarcated, and, as Whewell noted, it was mineralogy, geology's midwife, which was downgraded (Oldroyd and Albery, forthcoming). This was partly no doubt due to the abandonment of the mineralogically-based Wernerian system. The Wernerians had staked all on the claim that a proper mineralogy was the foundation of all 'geognosy'. With English geology's intense field involvement with sedimentary strata, and the discovery of the significance of fossils and of local conditions of deposition for understanding the meaning of strata, mineral lithology lost its explanatory power. During the two or three generations before the development of petrology in the mid-nineteenth century, a discipline divide was maintained between mineralogy and geology, which eighteenth-century investigators would have found extraordinary.

## The testimony of the rocks

I have been arguing that many aspects of landforms, fossils and minerals had been made into problems by investigation of strata, and were to be solved within the framework of new knowledge of strata. Indeed, investigation of strata became the focus of contemporary British Earth science.

Enumeration of the strata of which any country is composed, with an account of the changes they have undergone, and the operations that have been performed upon them, forms the true geological history of that country and an accumulation of many of these detached histories affords the best materials for a general history of the physical world. (W. Richardson, 1811: 368)

I shall now examine how the strata themselves were directly studied by the two clusters of investigators discussed in the previous chapter. The former might be called a lower class group, chiefly with provincial roots, professionally occupied with the Earth. The latter community was higher on the social and educational ladder, including many gentlemen amateurs, associated with metropolitan science, the universities and the Geological Society of London.

The opportunities, and problems, of surveyors and mines engineers sprang from their situation. On the one hand, they possessed often unrivalled local empirical knowledge of strata, the product of their 'acquired habit of observation', and their prime aim was to realize this knowledge as empirical, regional, stratigraphical geology. However, the previous generation of investigators into strata – Catcott, Michell, Cavendish and others – had hardly published their findings. Hence no established models were ready to hand within which these surveyors could work. This also suggests that men such as Michell had actually failed to find appropriate forms within which to integrate and articulate strata data. Furthermore, such problems were multiplied for the provincial consultants by lack of education, shortage of money and time, and other obstacles which further complicated the task of realizing knowledge into science.

The challenge they faced was not of constructing a deep philosophy of the strata within Earth history. For that was not their main ambition. Their interest lay in local geology, and for this they already possessed the requisite general working assumptions. By 1800 it was common knowledge that the strata were to be found in some kind of regular order, for which fossils and topography could serve as indices. The crux of their labours, however, was the struggle to articulate and communicate their knowledge. This rarely involved less than a long-drawn-out, tortured, craggy path to success. The tale of William Smith's fifteen years' agonizing on his *Map* is too well known to need retelling here (Bush, 1974; J. M. Eyles, 1969; Sheppard, 1914–22). John Farey clashed angrily with Greenough precisely over the issue of the form of an article submitted for publication in the *Transactions* of the Geological Society (1812) and with the Board of Agriculture over the organization of his *Derbyshire*. But Smith and Farey were fortunate in soldiering through to scientific expression. Trevor Ford has recently pointed out how much of White Watson's labours remained wrapped up in silence, in part because of his obvious difficulties in finding adequate media for representing what he knew. His one geological map is obscure. His famous inlaid sections of Derbyshire, employing samples of the rocks depicted, may have been one attempt to solve these difficulties (1960).

And of those who reached print, how many attained patterns of thought and forms of organizing their material which successfully communicated their stores of knowledge? For instance, Farey and Bakewell published numerous, lengthy prose accounts of the topography encountered and the strata crossed on their journeys. Essentially they were publishing their field notebooks. Clearly, this was more helpful than Catcott's or Cavendish's notebooks *never* being published. But it is dubious whether such straggling and disconnected reports really constituted the most effective way of communicating such information (particularly in journals like the *Philosophical magazine* and the *Annals of philosophy* which did not provide maps, diagrams or sections). It might have been more profitable to have stored up such data for more analytically organized regional surveys – indeed, like Farey's *Derbyshire* – but that is just what someone like Farey could not afford to do out of his own pocket.

In this respect, the noblest, but most signal, failure of the age was John Williams's *The natural history of the mineral kingdom* (1789). This was a mine of information – possibly more knowledge than any contemporary possessed – ineffectively presented (cf. Millar, 1810: Preface), as its slightly dated title may indicate. Williams compiled some competent chapters dealing with mining techniques, and presented minute local stratigraphical knowledge, for example of the Lothian coalfield (1789: pt i). But though he recognized that the strata lay in some kind of order, both vertically and geographically, he had not seen stratification as a heuristic key for presenting to his readers his mass of local knowledge on such topics as coal, limestone, sandstone etc. Different rocks were simply parcelled up into isolated chapters, in which Williams deposited all his accumulated data from different situations and locations, in a way redolent of earlier natural histories. The work lacked an organizing principle which would have made it more manageable, practical, predictive. It did not become popular, nor did it serve as the pattern for future works. Williams threw in a diatribe against the Huttonian theory of the Earth, and a dissertation on how America was first peopled by the Welsh, but neither volume contained a diagram, map, section or any other graphic representation. In short, the success or failure of descriptive stratigraphy depended above all upon attaining effective organizing principles of presentation. And the remarkable achievement of this group was their fertility in devising a variety of novel techniques for articulating and conveying such knowledge.

For those investigating a mining area, perpendicular sections taken from shafts could suggest the intellectual and visual key for arranging and presenting a body of knowledge. J. H. H. Holmes's *A treatise on the coal mines of Northumberland and Durham*

(1816) exploited mining sections to build up a comprehensive understanding of the structure of the north-east, as did N. J. Winch, using similar materials (1817). Westgarth Forster attempted to depict the geological structure of the same area pictorially, by constructing a gigantic notional vertical column of the strata from Cross Fell to the North Sea, encompassing the lead mining area and the north-eastern coalfield. The section was to scale, used consistent shading, and was accompanied by a physical map and an elaborate verbal key (1809; 1821).

Surveyors and practical miners increasingly sought to produce structural views of the strata of a locality. Verbal presentation of such research was helped by the founding of new scientific periodicals, such as the *Philosophical magazine*, publication in which obviated the expense and risk of raising a subscription. Such articles were particularly effective when local stratigraphical facts were keyed into the comprehensive view and terminology of the strata of England, as pre-eminently established by William Smith. Herein lay the unique importance of Smith's work.

Smith's great achievement lay in tracing more completely than any other contemporary the actual order of the English strata, especially for Secondary formations, in relation both to a generalized vertical succession and to their geographical outcrops. For Smith, belief in the order of the strata became an 'exemplar' for research, shaping his ambition, which was to trace that order for Britain. His sections show a more assured grasp than Smeaton, Michell or Whitehurst before him of the subtle differentiation of strata between the coal measures and the chalk. His account of their order was sophisticated enough to become the basis of all future work. Thus Farey, in several of his *Phil. mag.* articles, and then no less in his *Derbyshire*, routinely began by expounding Smith's order of strata as the intellectual, geological and geographical framework within which investigation of a particular area must proceed. Smith's terms for the strata also held for Farey the crystallizing force which the Wernerian terminology possessed for many members of the Geological Society of London (Farey, 1807: 108f.; 1811–15; 1815a & b; 1818).

But, as Rachel Bush (1974) and Martin Rudwick (1976) have argued, crucial to the success of articulating and communicating a mass of isolated local knowledge into the grasp of regional geological structure was the development of the stratigraphical map.[10] Not only could the map, through its elaborate conventions, be a thousand times more economical than verbal description. But because the map realized a *theory* of geological knowledge, it could also be interpretative and predictive. Late eighteenth-century French maps had been 'mineralogical', using spot symbols for deposits, not continuous shading for whole rock-masses (Rappaport, 1969). In their general way, the maps produced by Maton (1797), Aikin (1797) and Robert Bakewell (1813) began to embody mere stratigraphical ambitions, but without exactitude of local detail. Aikin and Bakewell were mainly concerned to divide up the country between areas of Primary and Secondary rocks. Smith's map of 1815 was not merely infinitely more detailed than anything which had appeared before. Above all, it embodied laborious thought and trials about what precisely a stratigraphical map should represent, and how it could be represented. In Smith's kind of map and memoir, meticulous stratigraphical observations found their ideal medium (Sheppard, 1914–22; Bush, 1974: ch. iv).

Taken together, Smith's account of the order of the English strata and his map presented a practical working basis for future descriptive stratigraphy – by Smith himself, with his own later succession of local maps, and by others. Between 1750 and 1820 professional surveyors made a huge contribution to British descriptive stratigraphy – both cumulatively and in resolving intellectual and practical problems which lay in its path. Later in the century, with the growing success of the Geological Society of London and the founding of the Geological Survey, the heroic days of the practical men were over.

In parallel to those men, the scientific elite was also cultivating stratigraphy. Its members were turning to geological fieldwork, out of a love of travelling, Romantic delight in Nature, and rejection of the artificiality of mere natural history. Many of the early members of the Geological Society of London began as mineralogists. But mineralogical laboratory work easily led beyond identifying particular specimens, into the broader structural concerns of geology. Mineralogists and chemists such as Charles Hatchett, Berger and William Allen, began to undertake field excursions. Discovery of the complexities of Nature then some-times created dissatisfaction with an essentially classificatory mineralogical approach, and thus led to geology. For instance, Arthur Aikin noted in the preface of his *Tour* (1797) that he made the journey

partly for amusement but principally as a supplement to the mineralogical studies of the author. From the perusal of books and the examination of cabinet specimens, I wished to proceed to the investigation not of minute detached fragments, but of masses of rock in their native beds; to observe with my own eyes the position and extent of the several strata, the order observed by nature in their arrangement and the gradual or more abrupt transitions of one species of rock into another. (1797: v–vi)

Furthermore, once in the field, it was found that Wernerian mineralogical classification was frequently belied by the actual order of rocks. Strata of what was believed Primary limestone, according to Werner's lithological criteria, were found for example to contain fossils; schist was found underlying granite. Nature or Werner had to yield. Stratigraphical position was increasingly seen as a more natural index than mineral characteristics in grasping the meaning of the rocks.[11]

To delineate the strata of the British Isles was a perfect research project for the Geological Society of London – as its *Geological inquiries* acknowledged. It epitomized the path-breaking, factual, scrupulous inquiry as recommended by de Saussure in his influential *Agenda* (*Phil. mag.*, 1799). It would be co-operative. It harnessed the energies of members in the provinces, under directive control from the centre. It was a tangible project, which avoided the contentious, religiously-entangling, fruitless subject of theories. It could be projected as of national utilitarian importance in an age when natural resources were being more highly prized, amid growing fears that known British coal and metal supplies would be exhausted. Lastly, it was an attractively practical enterprise at a time when foreign geological travel was hindered by the Napoleonic Wars.

And indeed, stratigraphy formed the main work of the Society in its first decade, as early volumes of the *Transactions* show. In each of the first three volumes, over half the articles – and these the most substantial – were stratigraphical.[12] George Bellas Greenough's endeavour to construct a geological map of the

British Isles was pre-eminently a Geological Society project, he being the first president.

Greenough himself toured Britain extensively to mark the strata, and corresponded on the subject with hundreds of fellow workers in the provinces, from miners to the titled. He showed throughout a pathological distrust of all explicit theorizing about the Earth, and believed that integrated stratigraphical knowledge, realized in map form, was the one concrete mode of progress for the science. At the Society's hub for a generation, Greenough digested all new studies on the British strata, and incorporated them with his own researches into his 1819 map, which shows superior accuracy and detail to Smith's (1819b; Bush, 1974: ch. vii).

It is worth noting, however, that the Society did not actually perform its stratigraphical work (or any of its other activities) like Solomon's House – the model seemingly set out in the *Geological inquiries*. The bulk of strata observations were in fact made by a small cadre of London and university based members, travelling in more distant parts of the country during their vacations. Because it was the work of a handful of activists, often doing their fieldwork in each other's company on terms of intimate friendship, meeting and corresponding frequently, the stratigraphy of the Geological Society shows remarkable homogeneity. Significantly, two of the most active early fieldworkers, William Phillips and W. D. Conybeare, teamed up to write a best-selling text, the *Outlines of the geology of England and Wales* (1822). The unity of word and illustration was achieved so well because the Society's curator, Thomas Webster, no mean geologist, engraved many of the pictorial representations for the *Transactions*. Also, because the Society's stratigraphy was the work of a much travelled core, it easily transcended the problems of an excessively localized 'mineral geography' approach which beset many of the professional surveyors, and could assimilate Continental investigations.

Indeed, the stratigraphy of the geological elite is quite distinct from that of the professional surveyors, for as well as descriptive local stratigraphy, they had wider ambitions. For instance, Arthur Aikin and Leonard Horner investigated the order and disruptions of the strata with an eye to evaluating the merits of rival conceptions of Earth history.[13] James Parkinson, William Buckland and

W. D. Conybeare fixed upon strata to unravel the problems of the historical revolutions of the Earth and life. The mapping techniques – reflecting an understanding of the complex history of the strata – shown by James Parkinson or Thomas Webster were markedly more sophisticated than those of Smith and Farey (Bush, 1974: chs v, vi). John Kidd in his *Essay* (1815) and Greenough in his *Critical examination* (1819a) analysed the very *concepts* of strata and stratification: was there a consensus on the meaning of such terms? Did their meaning necessarily depend upon a theory of physical deposition, or could it be neutral and descriptive (Greenough, 1819a: 123)? Inquiry of this kind – narrower than de Saussure's *Agenda* but broader than that of the practical workers – begat that characteristic English alliance of descriptive stratigraphy and directionalist interpretation of Earth history which flourished in the age of Sedgwick, Buckland, de la Beche and Murchison.

By the early nineteenth century, many of the traditionally separate threads of inquiry into the Earth had been reconstituted into geology. The newly dynamic study of landforms linked together Earth's past and present, surface and structure. Investigation into fossils and minerals was now being conceptualized not with respect to a Chain of Being but to the material history of the Earth, and especially in relation to the strata. Local natural history had been transformed into the regional stratigraphy. The 1820s were of course to make the history of life a more important component of the British geological outlook. And in the 1830s Lyell's work introduced – or reintroduced – geological dynamics. But the fundamental tradition of nineteenth-century British geology, an emphasis on plotting the strata *in situ* in the context of Earth history, was firmly established by 1815.

# 8

## The constitution of geology

### An eruption of theories

I tried to show in the last chapter how a steady, mainstream current was building up within the quickening flow of Earth science towards the end of the eighteenth century, gradually re-shaping the interpretative framework which gave coherence to expanding traditions of fieldwork. Erupting onto this scene in 1788 (and again in lengthier form in 1795) was James Hutton's *Theory of the earth*, which sparked off a decade of counter-theories and intense controversy. This intrusion did not stop, or reverse, the current of geological practice just described. But it did to some degree narrow its course, and in many respects deepen its channel, reconfirming the directions which inquiries were taking.

It is important to analyse this decade of theories, particularly Hutton's, in some detail, since it is a crucial moment in the development of British geology, but one which had frequently been misinterpreted. A powerful historiographical tradition believes that Hutton was 'perhaps the first student of the earth who may properly be called a geologist' (Gillispie, 1960: 293), and that 'prior to Hutton, geology did not exist' (McIntyre, 1963: 2). This case is argued on the grounds that Hutton was the first truly empirical fieldworker, and that the essence of his *Theory of the earth* was to be the first theory built on the basis of field observation. Sir Archibald Geikie thus deemed the word 'theory' in Hutton's title 'unfortunate', for it may have suggested 'still another repetition of the endless speculations as to the origin of things', whereas his *Theory* was really 'a vast storehouse of acute and accurate observation and luminous deduction' (1962: 295, 297). In a similar way, Gillispie has reckoned Hutton's *Theory* 'almost a misnomer', 'because of its resolute refusal to speculate on origins, and a rigorously empirical approach', which meant that it 'accorded so well with the facts' (1959: 46). For Tom-keieff, Hutton was original in being 'unbiassed by dogmatic pre-judices' (1947–9: 390), and Hagner has written: 'before James Hutton's "Theory of the Earth" appeared in 1788, natural philo-sophy was based more on speculation than on observation' (1963: 233). All such views emphasize the major role of Hutton in giving birth to the infant science. He was, in Bailey's words, 'the founder of modern geology' (1967).

The argument of this book suggests a different perspective. As I have tried to show, Hutton was not the first to devote himself to fieldwork. Rather, in this respect, his work was part of a rising tradition. Nor was his fieldwork the uniquely defining character-istic of Hutton's geology. This is in no respect to deny that Hutton was a superb field geologist: this is well known and it is unneces-sary to elaborate on his skills here.[1] It was not, however, by being a fieldworker that Hutton was unique, but by the distinctive intellectual framework within which he undertook it. Hutton's philosophy of the Earth was unique in its day, and arose from a unique milieu. This milieu was doubly special, in being Scottish, yet also distinct from the milieu of Scottish *geology*. For this reason, Hutton conceptualized the Earth in a language and within a set of scientific expectations alien from those of most contemp-orary investigation. This had several major consequences for the wider development of the science.

Firstly, in so far as Hutton's *Theory* provoked counter-attacks, these too, in answering Hutton on his own terms, had to be couched in a framework which looked just as foreign – and indeed archaic. For, significantly, Hutton and his counter-theorists were old men, who had matured in the intellectual world of mid-century. Hence a kind of self-moving eddy of theory and counter-theory was generated, outside the main current of the investiga-tion of the Earth. Although this body of theories showed much greater affinity with the ambitions of late eighteenth-century geology than with earlier clusters of theories, Hutton, de Luc and Kirwan were nevertheless rival ante-diluvian saurians locked in combat against each other, but equally destined to imminent extinction (Davies, 1969: ch. vi).

Secondly, because Hutton's theory was marginal to the main trend of the science, many investigators felt no need to engage directly with it, or at least only to grapple with those piecemeal transformations and dilutions introduced by his followers, such as John Playfair and Sir James Hall, who themselves were only too well aware of the need to update Hutton's geology in line with newer methodological norms. Many younger geologists responded to Hutton's theory by investigating fragments of it. In 1805 Greenough toured Scotland to check for himself the Neptunist and Plutonist theories against Hutton's own field evidence, finding Hutton almost as wanting as Werner when he examined the rocks personally (Rudwick, 1962). Nobody outside his personal circle became a whole-hearted convert to Hutton's geology as a *system*, though many could absorb elements of his views, such as his ideas on granite, or on denudation.

Thirdly, whatever the long-term importance of Hutton for the development of geology, his chief immediate impact was essentially negative. For Hutton was probing problems foreign to the concerns of most investigators. Furthermore the Huttonian controversy appeared to many a fruitless disgrace to the self-image of their science. It was raising questions which could not be resolved, and spectres which threatened the good name of Earth science in a period of counter-revolutionary intellectual illiberalism.

To argue this interpretation of Hutton as an outsider, who did not spring from the mainstream, and whose impact upon emerging traditions of Earth science was essentially oblique, I shall develop three main points in the next few pages. Firstly, Hutton's science of the Earth was in fact – and designedly so – deductive and deeply theoretical, embodying a total philosophy of Nature, guaranteed by an articulated epistemology of science. Secondly, Hutton's philosophy of the Earth incorporated fundamental conceptions, spelt out elsewhere in his writings, on physical and chemical issues, metaphysics and matter theory. Thirdly, precisely these two elements imparted to Hutton's *Theory* its uniqueness as a realization of the Earth, and enable us to explain features of Hutton's geo-philosophy which would otherwise remain mysterious.

Hutton saw himself as no mere empiricist. He pinned faith

upon rational, deductive methods. Some of his most important observations were made to confirm, not to decide, contentious points, indeed after he had actually written his *Theory*. 'All the granite I had ever seen when I wrote my *Theory of the earth*' was some 'at Peterhead and Aberdeen', 'and no more', notwithstanding that his interpretation of granite was so revolutionary and crucial to his theory (1795, i: 214). 'A philosopher, in his closet', by proper reasoning, could achieve a higher comprehension of Nature than the mere observer (even than de Saussure, in fact).[2] The progress of science depended upon the formulation of new theories in the mind at least as much as on amassing information (1795, i: 307).

Hutton's *Theory* was properly received as a comprehensive, deeply theoretical attempt to explain the Earth's history, its present and future as a self-sufficient economy of Nature, operating through universal causes. For Kirwan, Hutton's theories were '*a priori* conclusions unsupported by facts, and contradicted by all natural appearances' (1799: 482; Dean, 1973). De Luc recommended Hutton to use his eyes more: instead of constructing a 'hypothesis', 'you should have first attentively studied our present *continents*' (1790–1: 227). John Murray thought that although Hutton had the merit of 'boldness of conception', and succeeded in gratifying 'the imagination with the semblance of grandeur and design', these were 'the only merits of the theory', and constituted no great achievement, for composing theories was not difficult 'when the restraint of strict induction, is not imposed'. 'In appealing to the proof from induction we have found the phenomena of geology entirely at variance with its principles' (1802: 253–4). Contemporaries rightly recognized that Hutton's theory was something of a throwback. It had not discarded the dimensions of traditional theory, and was suffused with natural philosophy and theology instead of enshrining fieldwork pure and simple. Rather, Hutton had endowed the traditional theoretical ambitions with profoundly different *content*.

A brief summary of Hutton's own exposition of his *Theory* will in fact confirm how utterly he had articulated every aspect of the Earth within a total, integrated philosophy of the geocosm.[3] The *Theory* began from the postulate that actual climatic and terrestrial causes were continually, however gradually, destroying all land. This process, although destructive, was necessary, for only through continual disintegration of solid rock into new soil could sufficient new supplies of fertile soil be produced to support vegetable, and hence animal and human, life (1788: 214). Either Nature possessed only this single directional trend towards denudation, in which case the means, while in the short term maintaining the basis for organic life, would in time thwart the end, since all land would eventually decay and gravitate into the sea (1788: 215); *or* a wise, designing Deity had ordained some compensatory mechanism for regenerating land, and preserving that organic life which was the Earth's final cause (1788: 216–17). But all Nature pointed to such a Deity. Hence men should search Nature for processes for continually reconstituting land.

Examination of strata showed that they had been formed on the sea-bed, because of the presence of fossilized shells and organic remains, and the mud, silt and sand-like materials which composed them (1788: 218f.). But the detritus of denuded land gravitated continually, via rivers, to the sea-bed. Together, these two facts constituted moral proof that strata had in the past been formed at the bottom of the sea out of denudated materials, and were still being so formed. But how did this loose material consolidate on the sea-bed? Not by chemical precipitation out of a general solvent, *pace* the Neptunists, for silica and some calcareous rocks were, Hutton argued, insoluble in water (1788: 255f.). One single cause however must be responsible for consolidating all the strata, since rocks were found so intermixed. That cause must then be heat, which, aided by the sufficient pressure obtaining on the sea-bed, would fuse all rocks (1788: 229).

From the fusion of rocks by heat, it was possible to deduce that the Earth had a central fire. If the regeneration of the land was to be complete, some process must raise new strata being formed out of the sea (1788: 261f.). Such a deduction was supported by the marks of elevation borne by existing rocks – folding, fracturing, contortion, faults. This proved that the expansive forces of heat which had consolidated the newly created strata had also been responsible for their elevation. Central heat would likewise – on the principle of the economy of forces – account for earthquakes, volcanoes, and mineral and metallic veins (1788: 276). Such a natural economy of the degeneration and regeneration of the Earth's crust was a steady-state cycle, in which all continents, all landscape features, were secondary, because *forces* were primary (1788: 285f.). New continents were forever naturally created out of the debris of former ones, and, like the planets, the Earth might wheel on indefinitely with 'no vestige of a beginning, – no prospect of an end' (1788: 304).

Even this brief summary indicates a thinker with theoretical ambitions deeply distinct from his geological contemporaries, working within substantially different traditions of natural philosophy and theology. This is indeed so. For the milieux of Hutton's career were those of Enlightenment Paris, Leiden, and then, for most of his later life, Edinburgh. And above all, his intellectual context was *not that of Earth science*, but rather that of the synthetic philosophical ambitions of the Scottish Enlightenment (Playfair, 1805). Hutton's desire to produce a deductive theory, and the shape of theory he produced, followed from his specific personal situation in the Scottish Enlightenment. He worked within the matrix of problems pinpointed by David Hume, and his ideas followed parallel channels to those of his friend Adam Smith. His thought was as typical – which often means in many respects atypical – of the Scottish Enlightenment as were Hume's *Dialogues concerning natural religion* (1779) and Smith's *Essays* (1795). Hutton's ideas played an equivalent role in Scottish geology to that of Hume's and Smith's writings in their own respective fields.[4]

To some degree, Hutton's relations with Hume and Smith are clear from direct personal evidence (such as remains, given that Smith ordered Hutton to burn his papers on his death, and Hutton's papers have mainly been lost). Smith was one of Hume's close friends and much influenced by him. Hutton was a great crony of Smith, and no less indebted. Hume requested Smith to publish posthumously his *Dialogues concerning natural religion*, just as Hutton did act as Smith's literary executor, and published his *Essays*. To some degree, the relationship is apparent from common concerns in their writings. But there is also a more general bond between the three – that of shared self-images and life situations. They were candid, elitist cosmopolitan bachelors, with few formal professional or family ties, leading lives as philosophers in search of truth; ornaments to, but transcending and on occasion coming into collision with, their Scottish milieu.

Hutton operated broadly within those problems for an understanding of Nature which Hume had diagnosed: problems of epistemology, explored by the *Treatise of human nature* (1739); problems of content and interpretation posed by the *Dialogues* `concerning natural religion* (1779). He offered solutions for the physical and terrestrial world by building up a philosophical understanding parallel to that employed by Smith for solving the problems of the moral, social and economic realm.

Smith set out, in his *Wealth of nations* (1776), a view of the moral and political economy as the self-sustaining equilibrium of opposed human forces, which Hutton's theory parallels for the natural economy. In his *Essays*, Smith expounded the classic embodiment of that faith in a philosophical grasp of reality which pervaded the whole of the Scottish Enlightenment, and which gave uniqueness to Hutton's geology:

Philosophy is the science of the connecting principles of nature. Nature, after the largest experience that common observation can acquire, seems to abound with events which appear solitary and incoherent with all that go before them, which therefore disturb the easy movement of the imagination; which makes its ideas succeed each other, if one may say so, by irregular starts and sallies; and which thus tend, in some measure to introduce...confusions and distractions...Philosophy, by representing the invisible chains which bind together all those disjointed objects, endeavours to introduce order into all this chaos of jarring and discordant appearance. (1795: 20)

Hutton then, in his *Investigation of the principles of knowledge* (1794b), strove to give systematic ontological, epistemological and psychological expression to Smith's notion that philosophy was the science of the connecting principles of Nature which brought order to the intellectual world. His *Theory of the earth* embodied the conviction of the *Investigations* that physical truth could *only* be discovered and expressed when thought was formulated in the most universal, theoretical terms. True Earth science was the philosophy of the connecting principles of the geocosm.

For all knowledge proceeded *via* the senses (1794b: 132, 318–27). Hence, knowledge of the outside world was a construct of the active human mind, employing sense data (1794b: 1, 83). But how was man to distinguish true (i.e. scientific) knowledge, from false? To trust in *facts*, as in vulgar Baconianism or as advocated by Common Sense philosophy, was to cut the Gordian Knot (Playfair, 1805: 84; Hutton, 1794b, i: 243). Rather, man had to devise means to establish right thinking. The safeguard against scepticism and error was proper logical organization and critical testing of knowledge through the standards of internal consistency. At every point at which man assimilated, understood and tested new fragments of knowledge, he achieved a new consciousness of knowing. Knowledge thus proceeded in a hierarchy of self-consciousness, rising ever higher from immediate sense information, through perception and conception, to ideas, judgment, science, philosophy and finally religion (1794b, i: 507). Science was for Hutton knowledge of natural things and their material causes. Philosophy was a higher stage, the comprehension of Nature as an organized system, through Nature's general principles, which were final as well as material causes.

Hutton's was then an extremely internal conception of truth and, hence, of scientific knowledge: 'Truth is not a thing. . .truth is not a fact. . .Truth and falsehood are things which cannot properly be said to exist in nature, being only distinctions which take place in regard to the mind of man' (1794b, ii: 258, 279). Thus, truth is the 'consistency of our ideas' (1794b, i: 536) for 'order is not a thing. . .it is the action of mind' (1794b, i: 590). Hutton's was thus a highly elitist, anti-Baconian notion of science. Since truth lay neither in things, nor in fact, nor in common sense, not surprisingly 'every person is not qualified to read that Book of Nature'.[5]

For the proper method of testing the truth of ideas of Nature was to reduce them to ever more general abstract principles (1794b, ii: 285). If true, they would then fully explain, *a posteriori*, the maximum number of physical phenomena, consistently with all other such principles, and also correspond with our *a priori* notions of the teleological organization of the natural world.[6]

Thus natural philosophy should graduate into natural religion (1794b, ii: 112), which was the only, but perfect, form of religion. As a Deist, he conceived that the Deity's rational perfection entailed that He had created an entirely regular, self-sufficient system of Nature, operating by natural laws, unbroken by miracles, and unaugmented by Revelation. Miracles would have breached the wisdom of the system (1794b, i: 307). The only understanding of God's nature available, therefore, was the study of His works, i.e. philosophy. Hutton's Deism conditioned his view of Nature as much as other geologists' Christianity conditioned theirs. For if Nature sufficed the Deity for His ends, Nature's forces must be adequate to Nature's products. Thus Hutton's uniformitarianism presupposed that the same forces had always existed in Nature, working at the same intensity, maintaining a substantially identical globe. If this were the best of all possible worlds, ever to have been in other than its present perfect state would have signified imperfection. For if 'in studying this part of the economy of nature, we may perceive the most perfect wisdom in the actual constitution of things' (1795, ii: 89); then 'How can a philosopher, who is so much employed in contemplating the beauty of nature, the wisdom and goodness of Providence, allow himself to entertain such mean ideas of the system as to suppose, that, in the indefinite succession of time past, there has not been perfection in the works of nature?' (1795, ii: 222).

Hutton's metaphysical Deism, therefore, set up against the Mosaic few thousand years not so much a long time-scale as indefinite timelessness. In his Deism lay the rejection of many other characteristic concerns of eighteenth-century Earth science. There could not have been a Deluge, for if *necessary*, that would have implied a defect in the regular system, while if an *accident*, it could only be a pointless catastrophe in breach of sufficient reason.[7] Neither could Nature as a whole show progression or retrogression, for perfection is necessarily static. So Nature could provide no clues of its creation or consummation. Hutton's natural theology ruled out, just as traditional natural theology demanded, evidential theology. Here is no case of inductivism liberating geology from 'endless speculations as to the origins of things'. Hutton's nescience about origins *was* his cosmogony.

I have been trying to show how Hutton's epistemological commitments demanded that he wrote a *theory* of the Earth; and that for him such a theory necessarily comprised fundamental principles of Nature (i.e. not merely generalized observations), teleologically organized into a system. But what did Hutton believe such fundamental principles were? Here it is crucial to recognize how concepts developed by Hutton in his medical, chemical, physical and metaphysical treatises suffused his grasp of the Earth.[8]

The first such concept is the cycle, a natural economy of growth, decay and regeneration in time and process. Its intellectual origins for Hutton's philosophy are many. Cyclical notions of human history were common. Astronomy, with the planets wheeling in endless revolutions was a potent analogy (cf. 1788: Conclusion; Playfair, 1802: 440). The cycle for a Deist confirmed the sublime principle of sufficient reason (1788: 284). Hutton's own thesis at Leiden was entitled *De sanguine et circulatione in microcosmo* (1749).

Whatever the possible sources, these problems of assessing the nature of the system of the Earth's economy – was it like an organism? or a mechanism? in decay, stable or developing? – were those posed by Hume's *Dialogues concerning natural religion*. Hume had suggested the organic cycle as one possibility. A cyclical view expressed Hutton's conception of the Earth as more than the standard eighteenth-century machine: as an organism as well:

> But is this world to be considered thus merely as a machine, to last no longer than the parts retain their present position, their proper forms and qualities? Or may it not also be considered as an organized body? Such as has a constitution in which the necessary decay of the machine is naturally repaired, in the exertion of those productive powers by which it has been formed? (1788: 216)

So Hutton fixed his attention on the cycles of Nature, and especially on the most mystical and all-comprehending of all, the great cycle which subsumed the organic and the inorganic (Tomkeieff, 1947–9: 396–7). In this, the basic ingredients of heat, light, water, soil and fixed air maintained plants, which in turn produced vital air and phlogiston to support animal life. When animal life decayed, it renewed the stocks of fixed air, heat, water, soil and light, for the rational cycle to begin again. Hutton had invented a perpetual motion organism. Throughout Hutton's works, phlogiston's function was to provide continuity in the *cycle* of Nature which linked matter to life, vegetables and animals (1792: 171–270).

Hutton's other organizing principle was to view all phenomena not as ultimate facts but as resultants of conflicting forces, existing in equilibrium (Tomkeieff, 1947–9: 390). Hutton derived this from his own metaphysics of matter. Newton's conception of matter as unitary, inert, hard, impenetrable body could not adequately explain natural phenomena such as heat, expansion or resistance. There must, of course, be a principle of passivity. But so also must there be one of activity. Hence, Hutton deduced that we should not conceive of matter as 'material'. Rather, actual bodies constituted an equilibrium between two diametrically opposite forces, the traditional contractile, pulling, passive force reducible to gravity, and an expansive, active, weightless pushing power, reducible to heat (Schofield, 1970: 274).

Such speculative notions were in harmony with the 'Newtonian' philosophy of active powers which pervaded general matter theory in the course of the eighteenth century (Heimann and McGuire, 1971). But Hutton was remarkable in applying these ideas to the *Earth*, at a time when most geologists still tended to understand Newtonianism as entailing a stable, passive, static Earth. How did these ideas inform Hutton's geological theory?

Hutton's uniformitarianism was at one level methodological: 'We must read the transactions of time past, in the present state of natural bodies; and, for the reading of this character, we have nothing but the laws of nature' (1795, i: 373). It was also partly a commitment to a steady-state conception of the Earth, derived from his rationalist Deism. But no less was it a function of his speculative philosophy of Nature. The whole Earth was viewed as an organism, degenerating and regenerating itself in indefinite cycles, its materials – rocks, living creatures and water – endlessly being inter-converted in this 'active scene of life, death and circulation' (1795, i: 281). The steady state of the Earth was a macrocosmic equivalent of the equilibrium of forces in microcosmic body – Hutton often referred to the Earth as 'body' and as 'macrocosm'.

In this context we can begin to understand Hutton's Plutonism. Hutton explained the operations of the globe through two major forces: gravity – as was common in eighteenth-century theories – and heat, a newer conception seen by Hutton as the cause of consolidation (together, in fact, with gravity, which supplied the necessary pressure), of elevation, of faults and veins, of earth-quakes and volcanoes. Reducing forces to these polar opposites exemplifies Hutton's model comprehensive explanation. Further-more, these two forces, between them the key to the history of the Earth, in equilibrium together also constituted 'body' in Hutton's metaphysics. Thus, Hutton was obviously pleased to locate yet another internal consistency in his ideas, the parallelism between the forces the Deity employed in the metaphysical microcosm, and those in the terrestrial macrocosm. Moreover, for maintaining his central heat indefinitely at a constant intensity, Hutton once more invoked the natural cycle. The 'subterraneous fire' was constantly fed with new combustible rocks formed out of vegetable matter (1795, i: 243).

This cycle was endless, because heat was needed to consolidate and elevate rock, but decaying rock was needed to produce soil to support vegetation, and vegetable matter, when decayed, was needed to restoke the Earth's central heat *ad infinitum* (cf. Gerstner, 1968; 1971). This final example suggests how Hutton's geo-philosophy was inseparable from his total methodology of natural inquiry and general natural philosophy. In this matter, even Playfair would not follow his master, abandoning the sub-terraneous fire stoked by combustible matter, and seeking a justification for eternal heat in that grand source of natural philosophical speculation, the 'Queries' to Newton's *Opticks* (1802: 182). Murray in his *Comparative view* perceptively ex-posed both, and penetratingly attributed Hutton's error to his attempt to construct an impossible cycle (that of fire and heat) by analogy with true ones, e.g. the climatic (1802: 71–2).

Hutton's theory pinpointed the transformation of investigation of the Earth since the later seventeenth century. No less than for earlier theorists, Hutton saw the Earth designed as a habitat for man. Teleology was still a valid principle of inquiry. The Earth was rational, ordered, perfect, without blemish. There was no waste and no decay. There was no natural overall directional change – Hutton in effect agreed with Woodward, Whiston and Ray on *this* point. The Earth was like a machine or organism.

On the other hand, Hutton's notions of the *means* by which these natural ends were fulfilled were new, demonstrating the impact of Enlightenment conceptions. For Hutton, designed stability was maintained by active processes, central heat, earth-quakes and volcanoes and the gradual uplifting of newly forming strata from the sea-bed. All had undergone change, revolution and re-creation. Nature was active. The Earth was a product of time and Nature, not a *datum*. It was incalculably old.

The same Janus-faced quality informed Hutton's very conception of the science. He believed in the propriety – indeed the necessity – of writing a theory. No less than Burnet, he believed Reason pointed to natural truth. His explicit and intimate coupling of Earth history with ontology, epistemology, matter theory and physico-chemical thought links him more to previous generations than to his immediate successors. Indeed, Hutton clung to some of his obsolescent ideas – such as phlogiston – while his friends abandoned them. Yet, many aspects of his approach and method broke with the theoretical ambitions of the past and gave expression to ideas which contemporary geologists could share. He emphatically distinguished geology from revealed religion and human history. He demanded thorough fieldwork. He insisted that science ceased beyond the known order of Nature and experienced actual causes. His was a cosmogonical vision which denied traditional cosmogony; a teleological, speculative theory which eliminated all but actualistic evidence. Once dismembered, and mediated through the work of Hall and Playfair, many *particular* insights of Hutton became part of the fabric of nineteenth-century geology.

Hutton's theory – and similarly the comparable Earth philosophies of George Hoggart Toulmin (Davies, 1967) and Erasmus Darwin – drew two kinds of response. The younger generation who identified themselves with the specialist consciousness of mineralogy or geology looked coldly upon Hutton's ambition of a fully articulated theory, but, like the Edinburgh Academy of Physics, chose to test fragments of it – or, like Jameson, to attack specific aspects. Certain naturalists, however, chose to fight Hutton on his own terms and territory. Most prominent amongst these were Jean André de Luc and Richard Kirwan, though one might add John Williams, Philip Howard, and, in a rather delayed-action fashion, Joseph Townsend. Against Hutton's Deistic, uniformitarian, steady-state naturalism they championed direction-alist, catastrophist, Biblical theories of the Earth.

The responses of these men were not typical of the public face of Christian Earth scientists. For, during the course of the eighteenth century, a steady, though gradual, practical disengagement had been coming about between theology and the science of the Earth, marked by friendly *de facto* boundaries between the fields of discourse. This was partly due to the growth of narrower, specialist research on the Earth (such as local natural history, or lithology) which had little immediate relation to divinity. It was partly also an expression of Enlightenment Christian epistemological humility which warned against searching with precipitate innocence for a total system of truth to harmonize natural and revealed knowledge. 'Revelation and history are distinct things', wrote Sir Richard Joseph Sulivan (1794, v: 190). Or, as the *Critical review* observed, 'It would be improper to consider the history of the creation as related by Moses, too minutely' (1788: 35–40, 36). Theologians like Richard Watson, and reviewers, increasingly accepted liberal interpretation of Scripture (*Critical review*, 1788: 115) including of the Earth's age. Some concluded that the Bible was not fully inspired, or that textual problems remained unresolved. Others noted that the Bible was not intended to teach natural philosophy (Sulivan, 1794, ii: 263; Playfair, 1802: 477).

In any case, the Earth had never been such a popular emblem of natural theology as astronomy or the science of living creatures. Mainstream divinity was resilient and rich enough to accommodate whatever ideas about the earth its investigators were pressing at that time – whether the order of a functioning system, or the disorder of the Deluge. Provided that men of science were not contending against the *creation* of the Earth, theologians were unlikely to be seriously embarrassed.

Of course, during the eighteenth century the fields of Earth science and theology did not become totally separate. Yet a greater differentiation was forming than had characterized the age of Ray, Burnet and Whiston. The differentiation, however, was generally the product of *de facto* agreement; neither the cause nor the effect of 'warfare'. However, the virulent and comprehensive anti-Huttonian attacks of de Luc and Kirwan in particular broke this *détente*, and inaugurated that sporadic but prolonged atmosphere of 'warfare' between divinity and geology which so troubled the nineteenth century (Millhauser, 1954; Gillispie, 1959; White, 1876).

Hutton's assailants, like Hutton himself, significantly enough, were personally and geographically outside the mainstream of contemporary investigation. They too were old men, educated into an earlier culture which pursued natural philosophy in a broader, more connected context. Kirwan, as president of the Royal Irish Academy, was in effect doyen of the Irish scientific community, and may have felt obligation in his public position to uphold Christian values. Furthermore, he seems to have undergone some kind of deep personal religious experience late in life, and conceived a mission to confront the enemies of Christianity on all fronts, especially attacking Humean views on epistemology and miracles (1802c: 187f.; 1799: 4f.). De Luc had grown up in the midst of Genevan Calvinist culture, inheriting from his father a vocation of defending Christianity against the *philosophes*. His earliest works, written in French before he finally settled in England, were full-scale defences of Christian natural philosophy (1779).

In part, then, these counter-theories are explained by their authors. But very largely they are also the product of their time. For, within a year of the publication of Hutton's *Theory*, the French Revolution broke out. Conservative hysteria in Britain against the Revolution was quickly to blackball all speculative natural philosophy, all science derived from Enlightenment naturalism, all views of Earth history which seemed to assail Christianity (Garfinkle, 1955; F. K. Brown, 1961; Morrell, 1971b). Hutton's *Theory* was guilty on all scores – with or without the misrepresentations critics introduced in a fever of emotional revulsion. De Luc, Kirwan and Williams openly attacked Hutton's geo-philosophy for the 'atheism', 'infidelity', 'anarchy, confusion and misery' it sought to promote (Williams, 1789, i: lix). This is not to suggest that such men were abnormal bigots, or spent their lives in intellectual witch-hunting. De Luc prided himself throughout his career on the mildness and reasonableness of his vindications of Christian science. His main scientific work in his first decade in England had been chemical and meteorological; Kirwan had mainly been publishing in the field of chemistry and mineralogy. It is rather to point out how the conjuncture of the

*Theory of the earth* and the storming of the Bastille was for some an intellectual catastrophe – one which was critical in shaping the precise configuration of geology at the turn of the century.

These writers took their stand on the claim that a true theory of the Earth would be harmonious with Scripture. Kirwan in his *Geological essays* made 'no doubt of demonstrating' the truth of the Mosaic account (1799: 6), and de Luc constantly claimed that from a multitude of facts, such as the modernity of the present continents, 'we have the greatest character of veracity impressed upon the Book of Genesis' (1831: 233). De Luc believed that men must stand up and be counted in affirming the intimate connection between the Bible and the theory of the Earth. 'Revelation preserved in all its purity among one of these [ancient] nations is the true cause of the progress which mankind has made in the study of nature, and the only guide that has directed them in it.'[9]

Because de Luc and Kirwan sought to counter Hutton's geo-philosophy with a divine cosmogony, they were forced to grapple with all the traditional problems of exegesis and theodicy. How did God act in the world? De Luc insisted the Deluge was the necessary result of the laws of Nature, because that grounded the Deluge most securely in natural truth (1831: letters iii, iv). Kirwan on the other hand saw it as a miracle beyond man's understanding, because he wished to stress God's miraculous providence, and could find no possible natural mechanism for it (1799: 66). Then how was Scripture to be interpreted? For de Luc, the 'plain sense of the expression used in Genesis' was his guide: 'the literal sense is the truth', though he himself was assailed on more than one occasion by Kirwan for not taking Scripture in the plain literal sense (1831: 233f.).

De Luc's theory is worth examining in some detail, for it represents a crucial juncture. Hutton's *Theory* was self-consciously elitist. It offered affront to common sense. It flaunted the chasm between erroneous common consciousness and true philosophy. It had no aspirations to win over the general intelligent public. It acknowledged the unbridgeable divide between everyday perceptions of the Earth and a true geo-philosophy. De Luc's labours, however, were precisely the reverse: to *maintain* the century-old alliance between Scripture, Earth science and common perception.

He strove to offer a realization of Earth history, intelligible to the general readers of the *British critic* and the *Monthly review*, but also credible to the foremost scientists of the day.

De Luc believed that in the beginning God had created a chaotic fluid, which contained the chemical constituents of all future rocks. Precipitation gradually took place, so that first Primary and then Secondary strata were deposited (1793–4: 467f.). As this was occurring on the bed of the chaotic-fluid ocean, shells and skeletons of living creatures were gradually embedded in the newly forming strata. Thus, de Luc sought naturalistically to explain the origin of fossils, independently of the Deluge and miracles, and indeed, far anterior to the Deluge, for he recognized a high antiquity for the Earth itself. Meanwhile, as new strata were being formed, the level of the ocean was falling, and, in time, ante-diluvial land appeared (1831: 107). Then, when all the potential matter in the chaotic fluid had consolidated into an Earth similar to the present one, and God had brought forth the animal kingdom and mankind, quite suddenly an almost universal Deluge occurred, occasioned by the total collapse of all hitherto existing land (except some 'floating islands') (1831: 170ff.).

When the waters subsided, all previous land was now sea-bed, and all sea-bed had been uncovered to form the new, post-diluvial continents (1831: 186). For de Luc, this revolution solved one of the greatest technical problems plaguing eighteenth-century geology: how to account for fossils deeply embedded within the existing continents by an explanation consistent with Scripture. Since this Deluge, which de Luc dated for certain by the use of 'natural chronometers' not more than five or six thousand years ago (1831: 22), the creative process had entirely ceased – no new strata, no new fossils were forming on the sea-bed. Neither were there destructive forces. What little denudation had since occurred had rather been instrumental in creating barriers against further decay. The Earth was now stable and quiescent (1794: 589f.).

De Luc's theory adroitly compartmentalized Earth history into seven periods, comprising two utterly distinct phases. Like Woodward, he saw the present natural condition of the Earth absolutely stable – and for much the same Christian, natural theological reasons. Yet he also sought to give a naturalistic history of the

Earth, and recognized that strata evidence, fossils and landforms testified to a very long, complex and active process. His strategy, therefore, was to refer back the activity of Nature, both creative and destructive, to an era disjointed from ours. The present was characterized not only by completely new continents, but, above all, by a new order of terrestrial laws. 'By tracing back the action of actual causes we are enabled to ascertain with sufficient precision the time during which they have operated, or in other words the time which had elapsed since the birth of our continents' (1831: 81). De Luc's theory thus deftly integrated disturbing factual evidence within the presentation he desired of a rational philosophy and natural theology. In his elegant solution the watershed between the previously directional, active Nature and the present stable system was that great monument of Christian history, the Noachian Deluge.

De Luc's geology shared many of the preoccupations of the previous century. He was concerned with origins, and with the philosophy of Creation – though it is noteworthy that his physico-chemical speculations about the origins of the Earth were not elaborately expressed until Hutton had seemingly denied its origin. He was engrossed with the 'denudation dilemma' – yet this issue too had been sharpened by Hutton's apparently atheistic vision of an Earth undergoing ceaseless change. His pattern of Earth history, comprising a stable present age, the seventh, the era of man, preceded by the great revolutions of Christian history, followed the pattern of earlier theories.

Yet, within this schema, de Luc's views also register the considerable sophistications demanded by empirical observations since the time of Woodward – and he himself was a distinguished field-worker. He fully accepted a long time-scale, and recognized the original deposition of the strata had been extended and gradual. The Earth had seen great revolutions. Present strata were made up of the detritus of former. Life's history was long. Extinction was a reality. De Luc strove to realize a cosmogony which expressed the providentialism of Genesis in scientific form – a form which simultaneously imposed stringent limits upon, but also guaranteed, Christian Earth history. The miracles of instantaneous Creation had been liberalized into a lengthy developmental process. The Hexaëmeron had been rationalized into a directional, naturalistic Earth history. The sharp seventeenth-century divide between revolution and stasis had been transcended. De Luc was a bridge between seventeenth-century Biblical literalism and the actualistic providential directionalism of the age of Buckland and Sedgwick.

De Luc may have won the popular audience. But his Christian cosmogony seems to have been greeted with stony silence by the contemporary scientific community (cf. J. A. de Luc, 1831: Introduction). This was not, one supposes, because at heart they rejected such a scenario of a divine Earth history, but rather because in that age, mineralogists and geologists, with their growing specialist self-consciousness, were deeply chary of allowing themselves to become pawns in other people's battles – between Christianity and Naturalism, Britain and France. During the period of the Revolutionary and Napoleonic Wars, most Earth scientists, willy-nilly, narrowed their intellectual sights, and sought to safeguard their pursuit by resolutely defining it over and against theory, theology and all other contentious issues. The confident public missionary alliance of divinity and geology was not to be re-established until the 1820s, when, with intellectual liberalization, and with geology's growing social acceptability, a group of front-rank Anglican clerical geologists emerged – Buckland, Conybeare, Sedgwick, Whewell – formulating a more sophisticated comprehensive directionalism. Even so, this met with a frosty response from the geological coterie (cf. [Fitton], 1823) as did Lyell's re-introduction of 'theory' in 1830 (cf. Sedgwick, 1831; Page, 1963: 42).

## Geology established

Around the turn of the century, the term 'geology' increasingly came into currency, becoming in fact the recognized portmanteau term for the sciences of the Earth. Whereas the third edition of the *Encyclopaedia Britannica* (1797) made no mention of 'geology', the fourth (1810) carried a long article under that heading. In 1810 the *Philosophical magazine* for the first time included 'Geology' in its title, becoming the *Philosophical magazine, comprehending the various branches of science, the liberal and fine arts, geology, agriculture, manufactures and commerce.* Greenough later remarked that at the time of the founding of the Geological Society of London 'the term geology...was...little understood', and claimed with chauvinistic exaggeration that there was hardly anyone 'in the Kingdom beyond the pale of our little coterie who aspired to be thought a Geologist' (Greenough MSS). The use of the term indicates the graduation into scientific investigation of objects not before studied, or at least not studied from that particular perspective. It registers the conjoining of previously disparate methods and ideas, and also marks the exclusion of aspects of the Earth now no longer judged internal to the discipline. Geology, wrote John Hunter in 1793, 'has nothing to do with the original formation of the earth itself', but has only 'a connexion with the changes on the surface' (Owen, 1861, i: 299: cf. *Encyclopaedia Britannica*, 1797: 'Stones').

Constituting geology involved defining it. In part this was a matter of setting it over and against other sciences. 'Geology differs from cosmogony as a part from the whole' (*Encyclop. Brit.*, 1810: 'Geology'). For de Luc, the distinction between geology and natural history was that between a causal and a descriptive science (1831: 1). For Kirwan, mineralogy was the alphabet of Nature; geology studied those letters as articulated in Nature's book. In part it involved specifying the physical object of study. All definitions stressed that geology's business was with the Earth's *crust*, 'the internal structure of the earth, as far as we have been able to penetrate below the surface' (*Encyclop. Brit.*, 1810: 'Geology'); 'the natural history of the mineral beds' (Warburton, 1814); 'the different mineral substances that compose the crust of the earth' (*Rees Cyclopaedia*, 1819: 'Geology'). Furthermore, geology was a distinctive way of seeing the strata which made up the Earth's shell. It was to study them as part of the 'system of the earth' (*Geological inquiries*, 1808). In other words, geology was to be a science which focused on *relations between objects*, not merely on objects themselves. Geology examined the 'relative position and mode of formation', 'the relative structure and ages' of the rocks (*Rees Cyclop.*), the 'relations which the different constituent masses of the globe bear to each other' (Kirwan, 1799: Preface). Treating mineral masses as part of a 'system of the Earth' demanded the study of 'the changes which have taken place in these' (*Encyclop. Brit.*). Hence, it meant focusing on the 'laws of their changes' (*Geol. inq.*) which entailed that geology must be, as de Luc insisted, a *causal* science (1831: 2).

Central to all these definitions was an insistence on the precise limits of the new science. It was not the only science which dealt with the mineral kingdom – for such were mineralogy, chemistry and crystallography. Neither was it the only science which embraced the Earth, for so did geography and cosmogony. But it was the science which investigated the superficial features of the Earth's crust in their mutual relations, local, temporal, causal (W. Richardson, 1811: 368).

A further emphasis in all contemporary definitions is that geology was an empirical science of observation, distinguished from cosmogony with its speculative overtones. Hutton had argued that theory was the heart of scientific knowledge. Others like Erasmus Darwin staked a claim for the heuristic value of theory, particularly in the infant steps of a science (cf. 1791: 'Apology'). But the early nineteenth century produced a general onslaught against theory in geology, or rather the stance that geology and theory were mutually exclusive; that theorizing belonged to the prehistory of the science. For this is the moment when the historical myth was forged which I posed in my 'Introduction' as the central problem of this book. In these decades men such as Playfair, Fitton, Kidd and Greenough were establishing a historical *cordon sanitaire* to separate all previous Earth investigations – now dismissed as a mad nightmare of religious effusion, rhapsodic poetry and philosophical speculation – from the new, inductive, science of geology.

Am I by now in a position to explain why these early nineteenth-century investigators sought so resolutely to deny their past? I think the following diatribe of its secretary, John Ayrton Paris, to the Royal Geological Society of Cornwall will help to illuminate the attitude:

The fabulous and romantic age of geology may be said to have passed away; its disciples, no longer engaged in the support of whimsical theories direct all their attention to the discovery of facts, and to their application to purposes of extensive utility. The present era of geology may therefore, with much truth, be compared to that of its kindred pursuit, chemistry, when in the dawn of the sixteenth century it escaped for ever the trammels of alchemy, and assumed the rank and importance of an inductive science …The parallel may be carried still farther; for it must be confessed, that the benefits incidentally derived by chemistry from the visionary researches of the alchemists, and those which have been conferred upon the science of geology by the zealous but mistaken pursuits of its earlier disciples, are equally numerous, and alike important; the happy allusion of Lord Bacon to the fable of Aesop, respecting the former, may with much truth and force, be applied to the latter. 'They are like those husbandmen, who, in searching for a treasure supposed to be hidden in their land, have, by turning up and pulverizing the soil, rendered it fertile.'

Amongst the number of useful facts which have been thus brought to light, few perhaps, in the science of geology are more striking, none undoubtedly more useful, than those connected with the discovery of a certain regular order in the occurrence and association of the different mineral beds and strata, which form the surface of the globe. (1818: 168–9)

This is, as I shall explain, a revealing and key statement. For, to speak highly schematically, we have seen three periods which assailed visionary speculations and pinned great faith on factual observation, each backed by the same general 'Baconian' ideology. The first was the era of the founding of the Royal Society. The almost complete absence of fieldwork heretofore gives that particular attack its rational pertinence. Secondly, the period after the plethora of late seventeenth-century theories also produced an assault on hypothetical science. This was justified both in the name of Enlightenment empiricism, and also by the contrast between successful Newtonian 'inductivism' and the impasse reached by the age's theories of the Earth. But why did the third period, the early nineteenth century, witness such an overwhelming rejection of, and distancing from, earlier work? For flourishing traditions of fieldwork observation *had* been built up during the eighteenth century. These had – albeit, according to Paris, despite themselves, and almost unintentionally – succeeded in amassing data of a kind respected by the early nineteenth century, such as knowledge of strata.

In part this exaggerated dissociation of current inductive geology from a mythical 'theoretical' past was the geological community's attempt, in the face of enormous external pressures towards intellectual conformity in an age of counter-revolutionary turmoil, to convince society – and itself – of the untainted loyalty of its science (cf. Garfinkle, 1955; cf. F. W. Jones, 1953: 220f.). The religious Dissenters who formed a very substantial part of the Metropolitan community must have felt particularly vulnerable to blanket assaults against radicalism. Geology accommodated itself to its external environment by drawing its public intellectual boundaries very narrowly. That narrowness was part of a bargain which geologists in effect made with society at large. For the body of knowledge and working assumptions possessed by geologists were, after all, potentially radical and disturbing: a threat to contemporary common sense. The high antiquity of the Earth, its many revolutions, the impermanence of continents, the discovery of enormous fossil saurians, the confirmation of extinction, the gradual refrigeration of the Earth – such ideas were genuinely not welcome to the Christian cosmology of the Napoleonic War years. To achieve a social licence to pursue such studies, geologists had to realize their science in the hardest, most factual, terms, and temporarily to refrain from explicitly projecting this science as part of a wider cosmology. Tracing the strata of England confirmed the real world, was patriotic and possibly even useful. Whereas to stress the theoretical aspects of geology was bound to be unsettling; was, in a very literal sense, to take what had been the solid ground away from under people's feet. Looking over his shoulder towards France, the Rev. George Graydon concluded his work on the volcanic regions of North Italy by stating that theories were at best

to be considered as contributing but remotely to the more useful and serious objects of life. But when applied, as we know they have been too often, to excite and diffuse doubts of the most essential truths, and ultimately to sap the foundations of religion, and with it, of both private and public virtue, order and happiness, and indeed of the very existence of civil society, as too fatal modern experience has shown, it is not easy to say whether we shall be most struck with the vanity and presumption, the folly, or the wickedness of the attempt. (1791: 311)

Fortunately, the task of banishing theories was not too difficult. The major theories of the 1790s could be ritualistically dismissed as antiquated. Both Hutton and de Luc had espoused pre-Lavoisierian chemistry. De Luc had his 'pulvicles', Hutton his lumbering prose style. The theories were the work of Irishmen, Swiss and Scots. Scots 'feelosofy' and love of controversy were easily ridiculed. Hutton was a perfect victim for theory-bashing, being both a Deist and an overt defender of speculative cosmogonies. No 'insider' within the ambit of the English, metropolitan group broke rank and wrote a theory. In short, the dictate of prudence for geologists mindful of their precarious social position and their own peace of soul was to distance themselves from the possible wider ramifications of their science. At more propitious times the prefaces to the first volume of the *Transactions of the Geological Society of London*, or its *Geological inquiries*, might have been a triumphal fanfare. Instead, they amounted to a few evasive, low-profile pages, conspicuous chiefly for all the issues tactfully avoided.

But dissociation from theory was not merely a tactful social strategy. It also marked what can only be described as a positive sense of joyous liberation which accompanied consciousness of the birth of the new science, geology: a feeling that at last, and for the first time, the proper nature, content and method of Earth science had been revealed, exposing a millennial horizon of assured and fruitful progress in research for the *future*. John Playfair's review of the first volume of the *Transactions* of the Geological Society exuded an unprecedented certainty about the direction of future research, a detailed knowledge of the work in hand. Only a decade earlier, Playfair had been the champion of Hutton's deeply individual theory of the Earth. Now, he asserted that solitary research in the past had led to faulty theories, and argued for co-operative empiricism of a staggeringly 'vulgar' – un-Huttonian – kind:

All this tends to show the necessity of setting many hands to work, if we would obtain a just view of the laws which guide, and have guided, the phenomena of the mineral kingdom. For attaining this object nothing is of such consequence as the description of particular countries, and an accurate exposition of the facts which they exhibit. *Indeed, if the face of the earth were divided into districts, and accurately described we have no doubt that, from the comparison of these descriptions, the true theory of the earth would spontaneously emerge without any effort of genius or invention.* It would appear as an incontrovertible principle, about which all men, the moment that the facts were stated to them, must of necessity agree; and something would take place like what has happened to the opinions of philosophers concerning the origin of fountains. Instead of a hundred different theories, about which they disputed with never ending sophistry, there would be a few general maxims, in which all men of sense and information would uniformly acquiesce. (1811: 209) [My italics]

Such a statement is not in itself more or less audacious than Burnet's desire to chase a thought around the world. But while Burnet's was utopian, Playfair's was a practical research programme.

Armed with a research programme of this immediacy and clarity, no wonder early nineteenth-century geologists had no time for side-tracking theories. 'Maxims' would come automatically in due course. Little wonder also that they rejected the science's past. Herein lies the significance of John Ayrton Paris's view quoted earlier. For, as Paris saw it, previous labourers had at best *unconsciously stumbled* upon important truths. What the younger generation of geologists, looking back on their predecessors, found most puzzling and exasperating was exactly the aimlessness of previous inquiry: its lack of recognizable order, direction, common goals, research programmes. Because early nineteenth-century geologists were so certain of their future, they had absolutely no need – except as a whipping boy – for their past.

Of course, the newly proclaimed *independence* and *objectivity* of geology as an organized science involved in many respects a false consciousness. At innumerable levels, geology still deployed the methods and findings of other sciences and traded on the metaphors, analogies and assumptions of the intellectual world at large (cf. Rudwick, 1974). And, of course, geology projected the *values* of its devotees. We do not for example take Greenough at his word when he listed his solely scientific reasons for rejecting organic progressionism (1819a: 281f.).

Yet contemporary professions of independence and objectivity are not merely mystifications to trap the unsuspecting historian. For in the configuration of its own interests, skills and problem areas, the actual practice of geology in the early nineteen century was deeply shaped by those perceptions of independence as a science and objectivity which were the conditions of its birth. For example, as a result of conceiving geology as an autonomous discipline, British geologists of the first half of the nineteenth century operated with an extraordinary degree of indifference to larger geo-physical or cosmological issues, and in unparalleled isolation from the community of the physical sciences (Burchfield, 1975). Similarly, one element in British geologists' very tardy acceptance of the high antiquity of man surely lay in the ambiguities of their professional conceptions of 'recent times' and in the conventional boundaries drawn between geologists, archaeologists and antiquarians (Bynum, 1976). English geology's independent conception of the paramount self-sufficiency of field stratigraphy led later to a delay in taking up geomorphology, microscopic petrology and tectonics. Furthermore, before Buckland's venture into diluvialism and Lyell's embracing of *Principles*, nineteenth-century English geology was quite remarkably reticent about constructing a wider overt cosmology, uniting God, man, Nature and the Earth. There was no apostolic succession of Huttonian theory in England, even in its empiricized, Playfairian, form. When 'actualism' became a watchword in English geology, in the mouths of Scrope and Lyell in the 1820s, it had to arise out of their own needs and circumstances – for example, as philosophers of volcanic activity. Similarly, what became the most popular general account of the Earth in Britain in the 1810s tellingly emerged from totally outside the community of English geologists – the Frenchman Cuvier's so-called *Theory of the earth*, published by the Edinburgh professor, Robert Jameson.

The main endeavours in England in the early years of the century were to establish a working basis for cohesive, co-operative practice within the science. There was a particular urgency evident in books, articles and private correspondence to standardize terminology and typology. The science was being retarded by confused and ambiguous language: 'the general and indefinite meaning attached to the principal names given by the miners to the stones of most frequent occurrence in the Cornish mines has often been observed and regretted by mineralogists who have visited Cornwall' (Hawkins MSS). Numerous concrete proposals aimed to remedy this situation. The Geological Society of London set up a Committee of Nomenclature. But 'after a careful inquiry they have found that the present state of the science does not allow them to expect that a uniform nomenclature can be established' (Greenough MSS: Report, 10 July 1810). John Farey challenged this pessimism. He expressed a hope to Greenough in 1812 that, in conjunction with the Geological Society of London and the Wernerian Natural History Society, they could together devise an acceptable nomenclature (Farey, 1812). Formally, there was no solution. Indeed tension increased, since Wernerian terms became popular amongst the better educated, while practical men stood by time-honoured names. But even on this point, considerable progress had been made *de facto* by 1815, with native English terms, familiarized by Smith, holding their own on the Secondary formations, and Germanic terms, for the time being, triumphant on the more primitive rocks ([Fitton], 1818: 323).

Furthermore, there was considerable reconciliation on matters of method. From Whitehurst onwards, commitment grew to natural causation, and the uniformity and constancy of the terrestrial natural laws. Hutton and the Huttonians celebratedly took their stand on the very issue of uniform natural causation. To go beyond the present order of causation was to go beyond science.[10] But the movement towards natural causation was not a specifically Huttonian legacy so much as a broad prescription within the science. Even Kirwan set out similar statements of procedure:

> In the investigation of past facts dependent on natural causes, certain laws of reasoning should inviolably be adhered to. The first is that no effect shall be attributed to a cause whose *known* powers are inadequate to its production. The second is, that no cause should be adduced whose existence is not proved either by actual experience or approved testimony...The third is that no powers should be ascribed to an alleged cause but those that it is known by actual observation to possess in appropriated circumstances. (1799: 1–2)

Kirwan's statement is typical of the growing awareness of the need for a distinctive methodology for geology as a historical science. He grappled with the problem of the rocks' testimony and of *what constituted* actual causes. Greenough in his *Critical examination* (1819a: essays iii, vii) and Kidd in his *Essay* (1815: chs. xx–xxvii) similarly discussed the problems of method for a historical science. Across the spectrum of geological issues, public discussion debated practical and conceptual obstacles to the steady progress of research – the problems of colouring maps, of identifying fossils, of organizing publications. That many of these were relatively technical problems indicates how the on-going scientific enterprise was increasingly securely established.

The appearance of self-professed geological text-books well illustrates this trend. Text-books presuppose that working first principles are adequately established, and point to a popular audience seeking instruction. Text-books themselves define for the future the approved slant of the science. Playfair's *Illustrations of the Huttonian theory of the earth* (1802), though not a routine text-book, was an early example of geological vulgarization. Among the first text-books proper were Robert Jameson's *Treatise on geognosy* (1808), and de Luc's *An elementary treatise on geology* (1809). But neither was greatly successful. Jameson's total commitment to the Wernerian system and his counter-productive use of difficult foreign vocabulary spelt his failure. De Luc in turn was concerned to settle old scores against rival systems, and so was less effective in producing a simple general outline. The first genuinely popular text was Robert Bakewell's *An introduction to geology* (1813; 4 edns. by 1833) which to some extent became the model for later successes such as Conybeare and Phillips's *Outlines of the geology of England and Wales* (1822). As a professional surveyor, geological lecturer and writer, Bakewell knew what the public required. His work studiously avoided difficult terms and dryness. It offered clear explanations of phenomena, glossaries, a wide variety of diagrams, and illustrative sections (1813: 349f.). It gratified a taste for the picturesque and the Romantic. Bakewell aptly illustrated his text with British examples, and presented them engagingly in a first-person narrative.

His *Introduction* is revealing of the attenuated range of early nineteenth-century English geology, in what it omitted. There was very little discussion on the cosmic place of the Earth. Bakewell was not concerned with its origin, its pristine history. He stressed, however, that the Earth was very old, and that the testimony of rocks and fossils indicated complex and gradual processes throughout the Earth's history in repeated revolutions (chs. i, ii, p. 325). Bakewell offered a generalized description of the Earth's development, and of life, through a series of stages leading towards the present. But he used this mainly as a pedagogic device for ordering his account of strata and fossils. The reconstruction of Earth history was not central to his purpose. He showed little interest in fossils as an index of the development of life, conspicuously ignoring Cuvier and the significance of his work for Earth history and the animal economy (15f.). His concern was the structure and relationship of the rocks, in which context fossil history had its limited significance.

Equally absent was any attempt to relate geology to revealed religion. Bakewell's geology was not deeply keyed to the Bible.

The few cautious references to natural religion did not have a constitutive role in the book (328). Also missing was discussion of those laws of physics and chemistry which had occupied so important a place in seventeenth-century theories, and for Hutton and de Luc. These too were being excluded from the subject matter of geology. Newton had lost his relevance. Geology had been narrowed to the rocks, their position, their history. For Bakewell this exclusive focus allowed the science to proceed.

Lastly, theory was absent. Bakewell scouted mere empiricism and approved theory's heuristic value. But his idea of a useful theory was the low-order generalization which articulated specialized inquiry: a theory of the origin of basalt or the formation of soil, but not total causal explanation for the Earth itself (114f.). He addressed himself to aspects of Werner's and Hutton's ideas on particular problems, but made no general assessment of these or other theories beyond a perfunctory gesture at the close of the book (325) – which is in marked contrast to the lengthy evaluations of previous theories found in most eighteenth-century popularizations.

The most characteristic 'text-book' feature of Bakewell's *Introduction* is that it was not expounded as a cumulative argument, consecutive theory or train of research. It was organized into relatively self-contained units, each dealing with one particular geological problem or field of knowledge. His technique was a pre-digested stage-by-stage exposition. He first discussed structural features of the Earth (ch. i). He proceeded to the practical distinction of rocks into Primary, Secondary and Recent, listing and describing these formations *seriatim* (chs. ii–vii). Though he discussed the implications for Earth history of this division of rocks, he emphasized that no theoretical agreement had yet been reached as to their temporal order or the mode of their formation. Nevertheless, his classification could stand for utilitarian purposes – for the discovery of valuable deposits, and for the day-to-day business of locating and describing strata. Always for Bakewell priority lay with the practical tasks of geology as an engine for exploring and assimilating the structure and materials of the Earth.

Most other aspects of geology received some treatment – denudation, veins, earthquakes, volcanoes (ch. x). But Bakewell's prime discussion was of the stratigraphy of the British Isles (ch. xi). Though it seems that Bakewell knew little of Smith's account of the succession of English strata, he had a good general understanding of Primary, Transition and Secondary strata, which he set down in a sketch map, and discussed in detail for particular regions (facing p. 255). For Bakewell, constructing a British stratigraphy was the pay-off of his earlier discussions of rocks and structure.

William Phillips's *An outline of mineralogy and geology* (1815) is a similar work. It is in printed lecture form, having begun life as a course of public lectures at Tottenham. Significantly, it is subtitled, 'Intended for the use of those who may desire to become acquainted with the elements of those sciences, especially young persons' (1815). Phillips confessed his work was a compilation. The list of authors to whom he acknowledged debts illuminates his notion of the component parts of geology, for he had taken his mineralogy and chemistry from Arthur Aikin, his geological observations from the *Transactions* of the Geological Society of London, and his broad interpretation of the Earth from Jameson and Cuvier. He recommended beginners to buy ready-made mineral and geological collections (Preface). Phillips's *Outline* is the closest contemporary approximation to a Geological Society text-book.

Phillips made the commonplace distinction that mineralogy was the study of rocks in particular; geology, study of the 'earth in general' (2–3). He dismissed former theories as speculative cosmogonies, and underlined that geology could only progress 'by the patient investigation of facts'. It was 'an inquiry into the history and present state of the surface or crust of the globe'. 'Disclaiming all theory, it is my purpose to adhere to the legitimate objects of science as they have just been described, without exerting one pretension beyond them' (65–7).

Phillips rejected speculation about Creation, and the Huttonian–Wernerian debates, as beyond the science's capabilities. Rather, he presented descriptive generalizations about strata, fossils, landforms – generalizations which avoided theoretical controversy, but which confidently enunciated the established principles of the science, such as the general order of the strata or the great revolutions of the Earth (95–6). Phillips had no total causal theory, or entire history of the Earth, but rather a series of descriptions of particular rocks and a number of medium-level truths induced from observed facts.

Characteristically, Phillips had less to say on fossils and land-forms. Religion was also absent. He admitted a recent universal Deluge, but refused to speculate on its causes, nature or effects. He ended with a perfunctory natural theological peroration, but called that a 'digression' (192–3). Works such as Bakewell's and Phillips's drew the guidelines for normal geological science early in the nineteenth century.

This voluntary agreement to make geology the science of the pragmatic and the possible, to pose only questions which could be answered, or at least debated within the framework of the science, helped to secure its dynamic social success in the early part of the century. Within the geological community, defining and refining the object of geology gave unprecedented unity to investigations. Furthermore, growing consensus on method served to draw controversy *within* the science, where it would be conducted according to informal rules. It was thereby rationalized as an integral part of the science's development, not a threat to its very pursuit.

The first quarter of the century saw a marked absolute expansion of geological activity and publication. But beyond this, the transmission and diffusion of information and ideas became more speedy, assured and fruitful. More thorough and detailed criticisms were being made of major concepts and research both through fieldwork, in private correspondence and in print. Leading geologists debated each other's ideas and reviewed their books in the journals as never before.[11] Whereas late seventeenth-century naturalists had polemicized against each other from thoroughly incommensurable standpoints, geological controversy now became more fruitful. Publications show evidence of a growing momentum in deploying and integrating research. Public problem areas became clearly defined. Thus, for example, Leonard Horner dovetailed his fieldwork on the Malverns with his controversial alternative explanations of the contorted state of the strata, and of the origin of granite (1811). Jointly conducted fieldwork became more common. Very few early nineteenth-century geologists suffered the kind of intellectual isolation with which their predecessors had had to contend. The first sixty years of the nineteenth century were filled with controversy – between Neptunists and Vulcanists,

Uniformitarians and Catastrophists, 'Silurians' and 'Cambrians'. But these were all debates *within* geology. While throughout the period, the science's frontier of positive knowledge marched forward, from 1835 increasingly under the aegis of the Geological Survey (Bailey, 1952).

Furthermore, the science was achieving increased public recognition. Gentlemen and their ladies were going fossil collecting on their Sunday rambles.[12] Metaphors from geology began to appear in common use.[13] By the 1820s and 30s, geology had risen to become one of the sciences most pursued by the general public, most debated in literature, most feared by quarters of religious and humanistic orthodoxy (Gillispie, 1959: 185). Public interest witnessed the arrival of the science, but it also secured its future, being the source from which a continuous supply of talent for the future was drawn. For nineteenth-century English geology held out as the arena of the amateur. Some of its greatest exponents were won through the ranks of the interested general public. The science's prominence was maintained for half a century by being at the centre of public debate.

# Conclusion

This book has argued that the science of geology – or by implication any science, or similar intellectual enterprise – is a highly complex construct of formal and informal institutions, practices, standards, beliefs and facts, forged over time. The very notion, *geology*, is an artefact, not available in the mid-seventeenth century, but fashioned by the early nineteenth. The existence of geology is a *terminus ad quem* no less than a *terminus a quo*. To imply that particular discoveries or individuals – say, Hutton or Smith – created or founded the science would ignore all the structures which made their work possible. Charles Lyell captured public imagination by making the Temple of Jupiter Serapis the frontispiece of his *Principles of geology* (1830). But the geological phenomena of the Temple, and some of its significance, had been set out in the *Philosophical transactions* as early as 1757.[1] Many of the ideas of Lyell's *Principles* derived from Hutton's *Theory of the earth*. But the roots of Hutton's work lay in the philosophy of the Enlightenment and the massive contemporary travel literature which he had digested.

Through the latter part of the seventeenth and all the eighteenth century, facts were being established, and perspectives developed, disturbed and modified. By the early nineteenth century a clear concept of geology was coming into focus. But this was not – despite contemporary self-consciousness – a jettisoning of the past. It was the past bearing fruit. Kuhn is clearly correct to see the development of sciences involving restructuring from early incoherence towards higher integration (1963). But geology does not suggest this was a revolutionary process. Neither did the coherence of geology depend on achieving a paradigm, in Kuhn's stronger sense of an all-embracing *theory*. Early nineteenth-century geology possessed no such theory – indeed, it rejected the very idea. But it did possess a research programme, capable of being the dynamo of development within the science. A programme, not a theory, was best adapted to the needs of the time to serve as the basis for the expansion of routine nineteenth-century geology.

In the mid-seventeenth century there was no body of knowledge and ideas – no 'science of the Earth' – which stood clearly demarcated from the common beliefs of ordinary educated men. By the early nineteenth century, however, geology had its own conventions and perspectives which were distinct from common-sense wisdom. It required 'geological spectacles' to conceive the Earth as enormously old, as having undergone tremendous revolutions, or as continually active. It required geological training to understand the science's nomenclature, or how to read Smith's map. The point of George Bellas Greenough's *Critical examination* (1819a) was precisely to prove there was no common-sense, natural, neutral language of geology: all terms had conventional meanings endowed by their history and use. Ramsay of Ochtertyre found Hutton's theory typical of the mystification of 'modern philosophists, I will not call them philosophers' (Ramsay, 1966: 166). Robert Bakewell – who was usually sympathetic to popular understanding – patronizingly told the tale of the 'respectable Leicestershire farmer' who explained the shattered state of porphyritic rocks near Whitwick by recounting that 'all these rocks were rent asunder at the time of the crucifixion' (1813: 291). One of the pedagogic and literary devices used by James Parkinson in his *Organic remains of a former world* was to juxtapose innocent common sense against geological knowledge of fossils, strata and landforms (1804–11, i: letter 1).

This making of the science may be plotted as a succession of movements and stages. Its social dimension saw a development from the general scholar, to the all-round man of science, and finally to the specialist geologist. At first there were no scientific societies; then only general scientific societies; then eventually specialized geological societies. At first there were only metropolitan societies, but later English, Scottish and Irish provincial institutions. There was a shift from the almost exclusive involvement of the Oxbridge educated, of clerics and physicians, later broadening to include religious Dissenters, the lay middle classes and those who made their profession out of the Earth. All of these developments deepened and broadened the science's social catchment area.

Intellectually, we may trace a related series of stages and developments. In the mid-sevententh century, the Earth was investigated and conceptualized by a multiplicity of methods within disciplines with little common ground. By the early nineteenth century, study of strata, fossils, landforms and minerals could be integrated within geology. From within a cosmogony concerned to trace the origin of the universe, a geology had developed which investigated the history of the present crust: from Burnet's ambition of justifying the ways of God to man, to the geological textbook. The natural history of the Earth, with its timeless, universal grid of Creation, had yielded to the drive to understand the unique history of the materials of this particular Earth. The static had given way to the dynamic. Explanations in which miracles were central had yielded to more naturalistic viewpoints. Mother Earth had been objectified.

These shifts need to be explained. At one level, they are men's choices. To create science, and, *a fortiori*, to pursue Earth science, was a minority choice, which, albeit increasingly sanctioned by social values, remained contrary to prevalent expectations about what men should and would do. It remained a positive decision to study the natural rather than the civil history of a county; or to investigate fossils or minerals *in situ* rather than in the laboratory. Geology as a set of conventions, as an artefact, is the sum product of myriad 'deviant' choices of this kind; and is a different artefact in England from in Scotland, Germany, France or China; different in England in 1810 from what it was to be in 1840. These choices must be stressed to understand how the subject and configuration of a science is not a natural datum but a product of history.

Yet these choices were not random or made *in vacuo*. They had their context and logic. The choices were rationally made within defined socio-intellectual imperatives and constraints. To use an evolutionary metaphor: while social causes may not be helpful in explaining the original huge range of possible variations of approach open to those investigating the Earth, the logic of socio-intellectual situations will explain the selection of those possibilities which grew and throve. Social causes may only rather trivially illuminate the variety of types of study of the Earth which sprang up, or the gamut of interpretations proffered. But powerful socio-intellectual determinants explain why alchemical traditions died out, why collecting continued without ever becoming the hegemonic science; and why, on the other hand, geology flourished.

What then was the situational logic which pushed geology to develop and thrive? Socially, the period saw the continuous rise of science as a culture. Economic expansion and industrialization required better exploitation of natural resources, and hence more scientific knowledge of the Earth. They were also creating a profession of experts with geological knowledge. Perhaps more importantly, industrial growth and the scientific exploration of the Earth – home and abroad – stimulated by the growth of travelling, were changing *attitudes* to the physical environment. In some ways, as man conquered it, it became more friendly, less forbidding. In some ways, as society became more urban and industrial, the wilds of Nature grew in attraction. In some ways, as man became more powerful, his self-consciousness underlined the distinctions between man and Nature; between Nature before, and Nature after, man had conquered it. There is no linear and unequivocal trend in man's consciousness of his relationship to Nature. The point is that man's attitudes to the Earth were in the melting-pot, and geology was being shaped by, and was shaping these debates.

Similar pressures of logic shaped the science's ideas. Investigation of any particular problem uncovered data and suggested solutions which posed new ones. In particular, the complex history of the strata and their battery of fossils was scarcely dreamed of in the mid-seventeenth century, but, as it was brought to light, it gave new dimensions to other fields of investigation. The teleological directionalism of Christian universal history continually shaped theoretical attempts to ascertain the pattern of Earth history. Baconianism urged the science to continual accumulation of facts. Newtonian science injected massive confidence in the orderliness of Creation and hence the ability to comprehend the universe within the regular action of natural laws. The

Enlightenment strove to prove how active, regular, natural causes could effect what seventeenth-century scientists reserved for miracles. All these intellectual pressures gently led the investigation of the Earth towards geology.

On the broadest social plane, the making of geology had not yet, by the early nineteenth century, created a deeply disturbing impact. Geologists and laymen still believed that God had made the Earth for man's habitation, even if geology had changed ideas about when it had been created, and how it had adopted its present form. Up to the early nineteenth century, Earth science had not yet forced the educated public face-to-face with drastically alternative cosmologies. This is, indeed, one of the points of my argument. For seventeenth- and eighteenth-century investigation of the Earth had been parasitical upon other traditions of thought and outlooks, upon pre-existing views of reality. It could not generate alternative views until it had achieved some degree of independent existence as geology. It could not become geology until it had gone through a long process of being 'made'.

Only a bizarrely disembodied view of nineteenth-century geology would pretend that it inexorably discovered truths discomforting to traditional opinions, on the bland assumption that scientific truth necessarily prevails and that nineteenth-century geologists were perforce 'radical'. Almost all geologists combined commitment to the progress of the existing social order with their love of natural truth, and were convinced these reinforced each other (Gillispie, 1959: ch. vii).

It is, however, true that nineteenth-century geological discoveries were often profoundly disquieting; and that prominent social groups found them difficult to accept, even if geologists themselves generally engineered reconciliations between their discoveries and elements of older religious cosmologies.[2] Science in general, and geology more than most, undoubtedly assumed from religion some of the functions of guaranteeing cosmic and political order (Young, 1973). But it was becoming an order of Nature not of God, and one much more problematic in its comfort to men. Willy-nilly, the nineteenth-century geologist – however much a theologian *manqué* – had some relatively radical lessons for society. If seventeenth- and eighteenth-century study of the Earth were of no other significance, it was at least responsible for building up in early nineteenth-century British society a group whose own intellectual interests, career ambitions, techniques and commitments to professional standards, led them repeatedly to uncover truths painful to society, and sometimes to themselves.[3] Recent historians of science have been rightly concerned to underline the social and ideological interests and constraints which deflect scientists from their vaunted pursuit of objective truth. It may be worth re-emphasizing the symmetrical historical factors – social and intellectual constraints – which *do* make it scientists' business to uncover natural truths (Merton, 1967).

The gradual making of geology is, however, of significance in its own right, as an index and integral part of important currents changing European society and ideas. A vision of man, Nature and society which had been fundamentally degenerative, static, or at most cyclical during the Middle Ages and in early modern times was yielding to one which was dynamic, progressive, developmental and finally evolutionary. Geology was the product, the beneficiary of these shifts. But it was also the basis of them. Long before it was accepted that life, or man, had a fully extended history, or that the meaning of life and man were to be grasped through their historical development, the history of the Earth had been revealed and absorbed. The antiquity, history and development of the Earth underpinned the nineteenth-century temporalization of the science of its inhabitants.

Parallel to this, since early modern times the appeal to written authority – the wisdom of the Ancients, Classical scholarship, or Revealed religion – as the fount of truth on matters natural, historical and cosmic has been yielding to the appeal to Nature. In 1650 the truth about the Earth was largely still found in books. By 1800 it was in the rocks. The science of the Earth divested itself of the appeal to written testimony long before the science of human history. Geological 'pre-history' was accepted half a century before archaeological and anthropological pre-history, and was their distant midwife (Daniel, 1961). The new conceptual problems awakened in formulating the nature and methods of historical natural sciences were first encountered in geology (Whewell, 1837).

The same phenomenon has its social dimension. This period was beginning to witness that shift from humanist scholarship to science, which has steadily gathered momentum ever since. Britain is a particularly important instance because she largely lacked the public training of scientists for bureaucratic purposes so common in Continental Europe. The development of geology as a socially marginal activity in eighteenth-century Britain presents some important clues as to why men were beginning to opt for science, and to how science built up its characteristic forms.

# Notes

## INTRODUCTION

1. The Society's guide, *Geological inquiries* (1808), proposed that most members restrict themselves to collecting virgin facts. The Count de Bournon, one of the founder members, underlined that the Society's approach was 'celui qui supporte le plus de Baconisme'. Cited Rudwick, 1963: 335.
2. Raven (1950: 441) has pointedly remarked how Lyell is here reading back his own difficulties in harmonizing physics and divinity into the seventeenth century. Cf. Bartholomew (1973) and Porter (1976a).
3. See chapter 8 for why early nineteenth-century geologists constructed this 'catastrophist' view of their science's history.
4. Adams (1933a; 1933b). The shift in terms is twofold. Firstly, it is a shift from identifying the object being investigated (the 'terraqueous globe', the 'superficies of the Earth' etc.) towards stressing the subject discipline ('physical geography', the 'natural history of the Earth' etc.). Secondly, there is the coining and popularization of the term 'geology' itself. This term likewise pinpoints not the immediate natural object (the Earth) but the discipline, and furthermore registers a displacement in explanatory programme.
5. For other cultures' approaches to the Earth, see Blacker & Loewe (1975); for 'primitive' societies, M. Douglas (1970); for China, Needham (1959); and for the roots of Western cosmology, Jaki (1974).
6. Illuminating here is Pantin's distinction between 'restricted' and 'unrestricted' sciences, with geology among the latter (1968: ch. i).

## CHAPTER 1

1. Thus Childrey's sources and experience enabled him to write lengthily on certain counties (e.g. Cornwall) but reduced him to near silence on northern England, Wales and Scotland (1661). Before barometric and trigonometrical methods were developed in the latter part of the century, there were no reliable absolute or relative heights of hills. Thus, John Morton (1712: 18) noted the current belief that Northamptonshire was the highest county in England. The problems of establishing such elementary essentials as reliable heights were worsened by

local differences in the length of the mile (Plot, 1677: 'To the Reader'), and by the practice of calling the height of a hill the length of the walk from base to summit. Indicating relief on maps by 'mole-hills' was the commonest method in the seventeenth century. The next century pioneered hachuring. Contours were essentially a nineteenth-century development. See Bagrow (1964); D. G. Moir (1973); Skelton (1970).

2. Childrey (1661) is in this respect characteristically Janus-faced. Despite – or because of – his Baconianism, his sub-title promised an account of the 'Natural Rarities' of Britain including 'Diverse Secrets in Nature'. Later topographers, such as Plot and Leigh, remained credulous, but that is a different matter from Childrey's expressed *delight* in wonders.

3. The critical case is Thomas Bushell (Webster, 1975: 336–7; Gough, 1932). Once Bacon's secretary, Bushell claimed to be applying Baconian philosophy as mines prospector and surveyor. However, Bacon himself showed little interest in the mineral kingdom (though cf. Webster, 1975: 345–6); and Bushell's writings do not reveal specifically Bacon-ian ideas about the Earth and prospecting for minerals – merely a commitment to the utility of empirical knowledge. Bushell's Bacon-ianism was commercial coat-trailing.

4. Sir Thomas Browne will serve as a test case. He wrote about many aspects of the Earth – its Biblical history, its moral analogies, minerals and their symbolic meaning, the Deluge, changes of landforms – at times using Reason, his own observations, or written authority. But he never gathered his interests into a coherent whole. As an ardent Baconian empiricist his son Edward, by contrast, was eminently a man of the New Science, though his outlook was no more 'geological' than his father's. See Raven (1947: ch. xviii); Chalmers (1936); E. S. Merton (1949); Nathanson (1967).

5. A random glance at footnotes in subsequent books shows how far this basic information continued to be cited – frequently from abridgements of the *Phil. trans.* – throughout the eighteenth century. The pattern and timing of controversies (e.g. over fossils, or the Giant's Causeway) show how *Phil. trans.* articles could provoke further observation and ideas.

6. The most important productive mineral correspondence begun this way was that with Edward Browne, whose replies were published in *Phil. trans.* (e.g. 1669; 1670a; 1670b) and then independently in similar form as two books (1673; 1677).

7. William Molyneux distributed 'Quaeries' to secure the basic informa-tion for a contribution to Moses Pitt's *English atlas* (Hoppen, 1970: 200–1), but this came to nothing. *Phil. trans.* papers by members of the society on such problems include Bulkeley (1693), Foley (1694), W. King (1685), T. Molyneux (1697) and W. Molyneux (1683–4).

8. On this society I am much indebted to an unpublished paper of Anthony Turner. See also Hoppen (1970: 210).

9. The importance of endowed collections attached to institutions was that they were frequently enlarged – by purchase or bequest. Martin

Lister gave a valuable collection to the Ashmolean in 1685, Plot in 1691. Lhwyd presented his entire collection in 1705, Borlase his in 1755. Even the Anglophobe Conrad von Uffenbach grudgingly granted the quality of the Ashmolean collection (Quarrell and Quarrell, 1928: 26).

10. Ralph Thoresby visited London museums, finding Woodward's the most satisfying (Rev. J. Hunter, 1830, i: 340; ii: 158). For Lhwyd's visits to Beaumont's and Cole's collections, see Gunther (1945: 139f.).

11. Criteria of 'importance' must be flexible (Shapin and Thackray, 1974). I have considered someone 'important' who spent a considerable portion of his energies in studying the Earth, producing extensive observations or significant ideas. The following section will focus upon John Arbuthnot (1668–1735), John Aubrey (1626–97), John Beaumont (d. 1731), Thomas Burnet (1635–1715), Joshua Childrey (1623–70), William Cole (d. 1701), Nehemiah Grew (1641–1712), Edmond Halley (1656–1742), John Harris (1668–1719), William Hobbs (?d. 1743), Robert Hooke (1635–1703), John Keill (1671–1721), Charles Leigh (1667–1701), Edward Lhwyd (1660–1709), Martin Lister (1638–1712), Christopher Merret (1614–95), John Morton (1671–1726), William Nicolson (1655–1727), Robert Plot (1640–96), Abraham de la Pryme (1672–1704), John Ray (1627–1705), Tancred Robinson (d. 1748), Thomas Robinson (d. 1719), Richard Richardson (1663–1741), Sir Hans Sloane (1660–1753), Ralph Thoresby (1658–1725), John Webster (1610–82), William Whiston (1668–1752), Robert Wodrow (1679–1734) and John Woodward (1665–1728). My discussion is based on (a) MS sources such as the Sloane papers in the British Museum (for which see E. J. L. Scott, 1904), and the Ashmolean and Lister collections in the Bodleian; (b) printed biographical sources such as the *Dictionary of national biography* and the *Dictionary of scientific biography*; (c) printed letters, papers and diaries, above all Gunther (1925a; 1925b; 1939; 1945), Lankester (1848), Gunther (1928), Sharp (1937), Richardson (1835), Nichols (1809), and Rev. J. Hunter (1830; 1832). The fundamental survey of seventeenth-century British scientists remains R. K. Merton (1938), for evaluation of which see Rattansi (1972) and Webster (1974; 1975).

12. E.g. Plot, Lhwyd, Vernon, Whiston, Balfour, Sibbald, Ray, Keill, Lister, Burnet.

13. Some came from gentry backgrounds, e.g. Plot; others from humble origins – Ray was a blacksmith's son.

14. Those in Anglican Orders included Burnet (who had hopes of a bishopric), Nicolson (who became Bishop of Carlisle), Whiston, Webster, Morton, Ray, Childrey, Grew, Harris. Equally, none was what Leslie Stephen called a 'critical deist' (1876). Edmond Halley, however, was decidedly anti-clerical and was rumoured to be no Christian (Ronan, 1970).

15. Those also deeply involved in antiquarian studies include Thoresby, Woodward, de la Pryme, Morton, Leigh, Nicolson, Lhwyd. Schneer (1954) has highlighted the connections between antiquarianism and historical geology. For mythological studies see Manuel (1959), D. P.

Walker (1972) and D. C. Allen (1949); for chronology, see Manuel (1963), and Farrell (1973) who demonstrates the interconnectedness of an immense range of disciplines in Whiston's work.

16. This is not to deprecate the acute fieldwork by Ray or Beaumont. However, personal circumstances forced Ray to end fieldwork early in his career. Lister did much fieldwork in northern England (Carr, 1974) but his main published work reflected his classifying interests.

17. Woodward was ridiculed in his lifetime for the egotistic pride of forbidding anyone to touch his specimens. But this may also be a case of his developing sound rules for the maintenance of collections (Beattie, 1935: 210).

18. Exchange of specimens was a most important scientific and social function. When John Morton and Richard Richardson exchanged fossils they significantly used the key numbers of Lhwyd's *Lithophylacii* (1699) for identification (R. Richardson, 1835: 84–6). Hans Sloane, in his position of affluence, living in London, and secretary of the Royal Society, was particularly important in supplying provincials with news, specimens and books (Gunther, 1928: xv–xxiv). For Sloane and Richardson's exchanges see Brooks (1954: 127).

19. Other kinds of relations existed. Lhwyd, for example, was employed by the Earl of Peterborough to build up a fossil collection, while he was himself employing college servants, local quarrymen and labourers (his 'lithoscopists') to fossilize on his own behalf (Gunther, 1945: 133, 168, 177).

20. Contrast the later obviously fruitful practice of fieldwork in pairs or groups, e.g. by Catcott and fellow undergraduates in Oxford; Hutton, Clerk and Playfair; Maton and Hatchett; W. D. Conybeare and Buckland; Horne and Peach, etc.

21. For humanistic satire against world-makers and fossil collectors, see Beattie (1935: 210f.), Tuveson (1950) and Nokes (1975). For suspicion that fieldworkers were spies or revenue officers, see Gunther (1945: 355).

22. Generally, see Stoye (1952). For Continental travels, see Lister (1699) and Ray (1673). For Sloane's education abroad, see Brooks (1954: 42f.).

23. It would be redundant to list examples of the Continental works read by Englishmen: a glance at their correspondence or the footnotes in their books proves the point. Far more impressive is evidence – provided, e.g., by Lhwyd's letters – that they were *au fait* with work in progress overseas and eagerly awaiting the appearance of books such as Scheuchzer's *Itinera Alpina* and Lang's *Historia lapidum figuralium Helvetiae* (1708) (Gunther, 1945: 534f.). For Woodward's correspondence with Scheuchzer, see Jahn (1969). Woodward apparently claimed correspondence with over five hundred naturalists abroad (Gunther, 1945: 305). Many Continental Earth scientists visited England, including Scilla, J. J. Scheuchzer and Boccone.

## CHAPTER 2

1. The multiplicity of vying systems is revealed by the volumes any theorist felt obliged to refute. See, e.g., Stillingfleet (1662), Cudworth (1678), or Bentley (1692–3). Also see Colie (1957) and Mintz (1962). For the establishment of the Newtonian world-view, see Jacob (1971; 1974; 1976). In theory, Christianity provided just that set of working background beliefs which I am claiming was lacking, but in practice there was no agreement as to its precise implications for a philosophy of Nature. Liberal and literal interpretations of Scripture, voluntarist and rationalist concepts of God, providentialist and second cause notions of His operations were poles apart. Men looked to Nature to resolve these problems within Christianity just as much as they went to Nature to adjudicate between Aristotle and Plato, Boyle and Descartes. See Westfall (1973).

2. Classic examples would be the vacillations of Cambridge Platonists over Cartesianism (Colie, 1957) or Ray's uncertainties (Raven, 1950).

3. John Aubrey sought a copy of Plot's Queries for a correspondent in Somerset (Gunther, 1939: 41). Lhwyd's Queries were still available to Lewis Morris in the mid-eighteenth century.

4. For subterraneous forests, etc., see Beale (1666), R. Richardson (1697), de la Pryme (1700). For gigantic bones, see Th. Molyneux (1697; 1700), Tentzelius (1697), Lustkin (1701). For landslips, etc., see Wright (1668), Southwell (1683), Wm Molyneux (1697b).

5. Cf. Gunther (1945: 90). Lister (1674–5a: 323) reported that he had found in Yorkshire glossopetrae similar to those discovered by Ray in Malta. In his 'Brief directions' (1728: 93f.) Woodward gave copious instructions for noting the position of fossils.

6. The correlation of particular fossils with particular strata seemed to Woodward evidence of organic origin at the Deluge. But it could equally support the *lapides sui generis* view that particular rocks *bred* particular fossils.

7. Abraham de la Pryme solved this problem in a way favoured later by asserting that continents and oceans had changed places at the Deluge. This, however, was always thought to create severe problems with Scripture (1700).

8. Vestiges of this view continued. The opening example of Paley's *Natural theology* (1802), with its contrast between the historical significance of a watch and insignificance of a stone, depends on it.

9. Gunther (1945: 458). See also Richardson's letter to Sloane of 8 June 1702 in Richardson (1835: 53). In a letter to Ralph Thoresby, Woodward asserted that Richardson's claim that he believed all rocks fossiliferous misrepresented him (Rev. J. Hunter, 1832: 420f.). If so, that seems a significant modification of the position he took up in his *Essay* (1695).

10. As also, obviously, Steno in Italy. See V. A. Eyles (1958) and Scherz (1956).

11. For Morton's belief that his findings confirmed Woodward on this

point, see 1712: 98. Hauksbee thought he had empirically disproved the 'specific gravity' theory (1712: 541–4).

## CHAPTER 3

1. Cf. also Geikie (1962: 65f.), Zittel (1901: 29f.) and North (1934), who described Thomas Robinson's theory as 'so unscientific even in an unscientific age'. More sophisticated recent authors have excused theorists some of the more 'objectionable' aspects of these theories as forced upon them by public prejudice or clerical pressure. See Davies (1969: 10f.).

2. Of course they believed what they were conditioned to believe – but that is a different issue. For Halley, see Davies (1969: 12).

3. Though the systems of Hooke and Halley required them to postulate an Earth older than the 6000 years or so of a literal Bible reading, they did not believe it was hundreds of thousands, or millions, of years old. The bogey of Aristotelian eternalism or Lucretian chance loomed if *such* a long time-scale were proposed. For Hooke, see 1705: 335; for Halley, 1715: 296f., where he declares 6000 years the age of man not the Earth. Ray confessed that though Scripture indicated that God would set a period to the Earth, Nature showed no natural process of decay towards that end (1692). Generally, see Meyer (1951) and Toulmin and Goodfield (1967).

4. Burnet, for instance, believed that Scriptural events such as the Deluge gained scientific credibility if within the necessary laws of matter and motion rather than if miraculous, which was to 'cut the knot' (1684: 18). For Burnet, however, natural laws were no less providential than miracles. Erasmus Warren thought this was a dangerous limitation of God's – and human – free will. As faithful Newtonians, Woodward and Keill believed God enhanced by his miraculous activity (Davies, 1969: 73f.), but Arbuthnot berated Woodward for invoking miracles not specified in the Bible (1697: 10). These problems ran deep. Thus Whiston claimed that the Deluge was not 'a meer result from any necessary laws of mechanism', yet explained it by Halley's comet (1696: 132).

5. Ray defended teleological explanations of mountains: 'if we may allow Final Causes (and why may we not? What needs this hesitancy and dubitation in a thing that is clear?)' (1692: 253). But teleology itself was a problem area, not a solution. Thus, Hooke denied that fossils were *lusus naturae* on the ground that they would be a useless part of creation and 'Nature does nothing in vain' (1705: 318). But Plot believed they *did* have a purpose, 'to beautifie the World with those varieties', to which Ray answered that Nature makes all things ends in themselves, not merely to serve others (1692: 105).

6. Burnet (1684: Preface); Ray (1692: 56). The Creation and Deluge were themselves perfect test-cases for evidential theology, i.e. miracles and prophecies.

7. The main theories to be considered here are Arbuthnot (1697); Beaumont (1693); Burnet (1684; 1690; 1691; 1699); Hooke (1705); Keill

(1698; 1699); L.P. (1695); Ray (1691; 1692; 1693); Thomas Robinson (1694; 1696; 1709b); Warren (1690); Whiston (1696; 1698; 1700; 1714); Woodward (1695; 1714; 1726). Collier (1934) offers quite full summaries of many of these theories.

8. Of Robinson's theory, Lhwyd wrote 'Mr Woodward tells me it's not worth ye enquiring after' (Gunther, 1945: 233). Discussion of the Earth as an organism did not of course disappear – it is present in James Hutton (1785). But the *use* of organic analogies changed. For Robinson and Hobbs the Earth was a living animal, and hence mountains *were* warts, and tides the product of the Earth's heart-beat. Hutton by contrast used organic analogies in order to establish that the Earth was a complex, synergistic system of interrelating forces forming a self-regulating natural economy.

9. Thus Whiston explained Genesis's Creation account as neither 'a Real and Philosophical relation of the proper creation of all things' nor 'a meer mythological and mysterious reduction of the visible parts of it to six periods or divisions', but rather 'an historical Journal or Diary of the Mutations of the Chaos and of the visible Works of each Day' (1696: 87).

10. Burnet, Ray and Whiston – all in Orders – addressed themselves to the problem of the Earth's future. Significantly, those not in Orders who (however sincere their religious convictions) did not set themselves up chiefly as religious authors – such as Woodward, Halley and Hooke – did not speculate much on the Earth's future.

11. Burnet (1684: ch. xii). Still less did Burnet believe that present processes of nature were *creative*. Discussing mountains and valleys, rivers and seas, he asked rhetorically, 'By what regular Action of Nature can we suppose things first produced in this Posture and Form?' Earthquakes could destroy, but 'how incompetent such causes are to bring the Earth into that form and condition we are now in' (29–30).

12. Recent studies of Newtonian matter theory have rightly emphasized Newton's own use of 'active principles' (e.g. Heimann, 1973). However, 'Newtonianism' as actually applied to the Earth stressed the passivity of the globe and the activity of God. See Burtt (1925: ch. vii); Jacob (1969); Guerlac and Jacob (1969).

13. Though natural forces for change came to be seen as feeble and inefficacious. Ray wrote of the 'leisurely proceedings of Nature' (1692: 50). Woodward thought the forces of the Earth worked 'silently, and insensibly, and, which is the constant Method of Nature, with all imaginable Benignity and Gentleness' (1695: 227).

14. See also Ray (1693: 202f.), who, however, weighed up similar arguments for a universal Deluge and found them no more conclusive than the arguments against.

15. For Hooke, see Davies (1964) and Turner (1974). For Halley, 1692; for Ray, 1692: 66f. By the time he wrote the *Three physico-theological discourses* Ray had changed his mind, thinking that the shift in the centre of gravity would only produce a partial Deluge (1693: 118). Woodward also denied the shift (1695: 51).

16. Its theological sources did not include works published since the

original sermon, illustrating Ray's enforced switch from worshipping God as a parson to worshipping Him as a naturalist. For change in landforms, see 1692: 1–55; for the Deluge, 56–131; for the Creation, 150–70. See also Raven (1950: 432).

17. John Woodward (1695) first provoked L.P. (1695), Robinson (1696) and Arbuthnot (1697), against which Harris (1697) was a defence – to say nothing of the controversies which Woodward stirred up on the Continent, for instance with Dr Camerarius of Tübingen. In the same way, Whiston's *New theory* (1696) was attacked in 1698 by Keill, to which Whiston replied in his of 1698. Keill returned to the charge in 1699, and Whiston again defended himself in 1700.

## CHAPTER 4

1. Gillispie (1951) and Greene (1959) skip periods of the eighteenth century. Raven (1954), Rudwick (1972) and Davies (1969) subscribe in different ways to the 'decline' theory. Davies may well be right to argue (96) 'For fifty years after 1705 not one British scholar made a contribution of any significance to the literature of geomorphology', but when he complains that in this period 'from the writings of men of the standing of Hooke, Newton and Ray' the historian must pass to 'scraps of material culled from the publications of untutored travellers, second-rate topographers, and long-forgotten clerics', he misses the long-term significance of *such* men now becoming greatly involved in Earth science.

2. Mason's travel diaries are in the Cambridge University Library, MS Add. 7762. One who acknowledged Mason's help in discussing issues was the Rev. William Smith (1745: 153). For Michell, see Geikie (1918), and his earthquake work (1760). Mason and Michell were both interested in mining. Similarly, Alexander Catcott prospected in the Mendips in the 1760s.

3. E.g. Withering (1782); Keir (1776); Wedgwood (1783; 1790). For Lunar Society correspondence on mineral analysis cf. McKie and Robinson (1970); Zeman (1950); and Schofield (1963b).

4. I am deeply indebted here to a private reading of an unpublished manuscript by Dr Dennis Dean which analyses precisely the growth of geology seen as a Romantic mode of the grasp of landscape. See also Nicolson (1959).

5. Burnet's denial of original sin and his allegorizing of the Genesis cosmopoeia typifies a mind rationalizing towards Deism (Lovejoy, 1948). Hostility aroused by his *Archaeologia philosophiae* probably cost him a bishopric. Forty years later Stukeley accused Martin Folkes, President of the Royal Society, of spreading infidelity – 'when any mention is made of Moses, of the deluge, of religion, Scripture, etc, it is generally received with a loud laugh' (Piggott, 1950: 143).

6. For decline in belief in *prisca theologia* and *prisca sapientia* see Manuel (1959). Erasmus Darwin used the shell of such beliefs as a mock-heroic apparatus in his *Temple of nature* (1803). Those philosophical systems for which ancient wisdom remained central (e.g. Swedenborgianism)

were becoming demarcated from what was acceptable as good science (Hirst, 1964). For Toulmin, see 1785: 7f.; for Hutton, 1795, i: 187.

7. On the Bible not being fully inspired, see Sulivan (1794, ii: 209f., 237f.); and on not teaching physical truth, see Playfair (1802: 125f.).

8. For Buffon, see 'De la nature, première vue' (1749, trans. W. Smellie, 1785, vi: 249): 'Nature is herself a work perpetually alive, an active and never ceasing operator, who knows how to employ every material, and, though always labouring on the same invariable plan, her power, instead of being lessened, is perfectly inexhaustible. Time, space and matter are her means, the universe her object; motion and life her end'. For Playfair, see 1802: 117.

## CHAPTER 5

1. In two revealing instances a specialist wrote a section on the local natural history of the mineral kingdom within a general topographical work by another author: James Keir, in Stebbing Shaw (1798–1801), and Samuel Dale, in S. Taylor (1730).

2. Cf. for instance Mendes da Costa (1747); Henry Baker (1747); Lyttleton (1748); D. E. Baker (1748); Henry Baker (1749); Mendes da Costa (1749); Joshua Platt (1750); Henry Baker (1753); Brander (1754); Parsons (1755); Mendes da Costa (1757b); Parsons (1757); Platt (1758) for a representative decade of fossil studies in the *Phil. trans.*

3. For printed accounts of Mendes da Costa's activities see Nichols (1817–58), iv: 84f., 233f., 456f., 516f., 775f., 761f. His MS correspondence in the British Museum (Add. MS 28534–28544), upon which Dr A. Keller is working, witnesses his enormous private and semi-public network of British and foreign correspondents. Though Da Costa's main interest was collecting, this spilled over into broader concerns with the significance of these objects for a philosophy of the Earth (see e.g. Nichols, 1817–58, iv: 778).

4. The following section is largely built upon Catcott's MS diaries and travels from the 1740s to the 1760s (Bristol Central Library, MS B 6495) and his *Treatise on the Deluge* (1761). Cf. also the extensive discussion in Neve and Porter (1977).

5. Holloway was engaged for many years on a translation of Woodward's *Naturalis historiae telluris*, which appeared in 1726. For Holloway's correspondence with Woodward, see Bodleian Library, MS Gough, Wales 8.

6. There is an important attempt by Cavendish, possibly with Michell's help, to trace strata across the country (Harvey, 1971: 321f.) which begins: 'The uppermost strata across the island are the chalk and the strata formed from its ruins and under that lies a thick stratum of clay and under that at least occasionally formed a band occupying all most all the S.E. Part of the island.'

7. The identification of strata by Catcott, Cavendish and Michell bears comparison in competence to that done on the Continent by Lehmann, Pallas, Arduino etc. Continental naturalists pioneered the sharp distinctions between crystalline and sedimentary rocks, and their relative

positions, under the terminology of 'Primary', and 'Secondary'. The terrain *they* covered gave English investigators less cause to make such distinctions.

8. Hamilton's works are saturated with contempt for Italian Catholic 'superstition'. Strange cited the Senecan tag 'naturalem causam quaerimus et assiduam, non raram et fortuitam', which the like-minded John Playfair was later to use as the epigraph to his *Illustrations* (1802). Hervey's Christianity was an Enlightened benevolist Optimism seemingly lacking all dogma (Fothergill, 1974).

9. He advocated greater strata study for economic reasons as well (1778: 144). He announced his intention of depositing Derbyshire strata specimens in the British Museum (1778: 145).

## CHAPTER 6

1. From a transcript in possession of M. J. S. Rudwick.

2. The main published material on the society is R. S. Watson (1897). My account is largely based on research on the society's archives, especially the minute books for the first thirty years, brought together by an early secretary, Anthony Hedley, in a series of *Reports*. Cf. Porter (1973b: 329–30).

3. For the early years of the society, see Todd (1955–64). My account is mainly built on study of the society's archives, particularly the early minute books.

4. The Earl of Buchan called Williams 'an illiterate miner', though a man of great practical skills: MS note in John Williams (1787), National Library of Scotland copy.

5. Wilson Lowry, Thomas Webster and James Sowerby were engravers, artists and draftsmen; Richard Phillips, Arthur Aikin, William Pepys, Alexander Marcet and Richard Chevenix were all chemists; William Phillips and Alexander Tilloch were publishers. William Babington, James Franck, James Laird, James Parkinson and Richard Knight were practising medical men; William Allen and David Ricardo were businessmen.

6. Greenough's travels included a prolonged visit between 1799 and 1801 to Germany and central Europe; in 1801 to Cornwall; in 1802 to the West Country, France, Switzerland and Italy; in 1803 to Italy, Sicily and Malta; in 1805 to Scotland, Ireland and the Peak District; in 1806 to Ireland. Sir James Hall had travelled extensively throughout Europe in the 1780s and 1790s. See 'Tours abroad, 9 May 1783 to 14th April 1784' (National Library of Scotland, MS 6324); 'Journal 1791' (National Library of Scotland, MS 6329).

7. For Farey's attacks on Greenough, the Geological Society, and their Wernerian terminology, see 1813; 1815a; 1815b; 1819; 1820. Cf. 'An Engineer' (1815: 117), criticizing Thomas Thomson of the Wernerian Natural History Society, claiming he must 'submit to be told by practical colliers'.

8. For reports of whose activities see *Phil. Mag.* (1799, iii: 318; 1800, iv: 369–72; 1802, xii: 284, 364); Woodward (1907: 7f.); and the minute

book of the society (in the Mineralogical Dept. of the British Museum, Natural History).

9. 'The papers submitted to the public in this volume [i.e. vol. iii of the *Transactions*] will shew that the authors have continued to act upon the principle of avoiding the discussion of theories which are too systematic and extensive and of treating geology principally as a science of observation.' Note in the 1816 Report of the Council of the Geological Society of London (in Greenough MSS).

10. For the society, see its published *Memoirs* and the minute books in the Edinburgh University Library.

11. For the Academy of Physics cf. a letter to Henry Brougham, 13 February 1798, in its Correspondence Book. For the *Edinburgh review*, cf. 1802: 201–15; and 1803: 337–48; p. 348: 'among all the wonders that Geology presents to our view we think the most unaccountable is the zeal and the confidence of the theorists'.

12. This is patently so of Hutton, but it is also true, for example, of John Walker. His 'Occasional Remarks' sets out a full philosophical and moral framework within which science was to be undertaken; briefly, Moderate religion, Common Sense philosophy and Baconian natural history. The Academy of Physics likewise defined its own religious, philosophical and moral stances as a framework for natural philosophy. 'Extracts from the Minutes of the Academy of Physics', 7 September 1797 (MS in the Royal Society of Edinburgh).

13. See letters of John Williams inserted at the end of the Earl of Buchan's copy of Williams (1787) in National Library of Scotland, soliciting Buchan's aid with mining and quarrying projects.

14. Cf. an exchange between Thomas Allan and Wm Thomson. Allan wrote 'as a mineralogist or a practical geologist Hutton was by many degrees inferior to Werner; but as a philosopher he was infinitely his superior'. Thomson replied, 'I wish Mr Allan had here stated what he means by a *philosopher*. As far as the inductive philosophy of Bacon is concerned the term cannot be applied to Hutton's speculations at all' (T. Allan, 1814: 111).

CHAPTER 7

1. Hamilton, 1771: 2. 'The operations of nature are slow' (Hamilton, 1770: 4). Volcanoes grew 'by degrees' (1769: 19). Sulivan made the same point: 'Nature is slow in her operations, She proceeds by insensible gradations' (1794, ii: 2).

2. There was a fierce debate whether these were genuine volcanoes, or merely vitrified forts, as Williams argued (1787).

3. Bush (1974: ch. vi) has analysed the acceptance by Farey, Richardson, Parkinson and Webster of massive denudation. Though William Richardson was opposed to Hutton, he could still accept massive denudation (1808).

4. The manuscript notes of his lectures in the Edinburgh University Library (particularly the series DC 2, 17–21, vol. ii, p. 163f., *c.* 1784)

contain even stronger attacks on volcanism than are printed in Scott (1966: 202f.).

5. Revealing here is Parkinson's *Organic remains*, the first two volumes of which (1804; 1808) are self-confessedly planless (see Preface to vol. ii). In the Preface to the third volume (1811), however, he expressed for the first time an indebtedness to Cuvier and Lamarck for bringing new vision and classification to the study.

6. Thus Thomas Tredgold's complaints against the Wernerians, that they performed their stratigraphy not as geologists but as mineralogists, 'almost as if the rock occurred in detached patches – seldom describing it as a continued stratum; and instead of attempting to show the structure of the country...by maps and sections, the Wernerians content themselves with giving a string of technical terms connected by expressions which are scarcely to be understood' (1818: 37).

7. E.g. W. Maton (1797); R. Jameson (1798; 1800); J. MacCulloch (1811a; 1811b; 1819); Berger (1814); Kidd (1814).

8. See More (1782); Withering (1782); Watt (1804); Keir (1776); and generally C. S. Smith (1969). See also the report of the January meeting of the Literary and Philosophical Society of Newcastle, 1809 (in Anthony Hedley, Reports, ii): 'Mr Cookson produced a piece of glass which had been allowed to cool very slowly and so exhibited the opacity and other properties of a crystallized stony body. In illustration of this phenomenon, Dr Townson requested that an extract might be read from Sir James Hall's paper on Whin and Lava in the *Edinburgh Phil. Trans.* – This phenomenon has since been exhibited on rather too large a scale by the falling in of the dome of the furnace on Mr Cookson's bottle house in the Close. Some tons of the melted glass had in consequence of the accident an opportunity of cooling gradually and exhibited all the varieties between perfect glass and an opake crystallized substance.'

9. Sir James Hall (1794; 1798; 1805). Good accounts of Hall's experiments are V. A. Eyles (1961; 1963). See Hall's MS of his experiments (National Library of Scotland, 5019) in which he noted Black's as well as Hutton's opposition to his experiments.

10. Other forms of visualization were also pioneered, such as the conceptual block diagram of faulting, by Farey, presented in his *Derbyshire* (1811–15).

11. See T. Allan (1813); or, making the same point, a letter from William Wollaston to Thomas Allan, 25 September 1811, on the order of strata at St Michael's Mount (National Library of Scotland, MS 583, no. 749); and W. D. Conybeare to G. B. Greenough, 4 December 1811 (Greenough MSS).

12. The first volume of the *Trans. Geol. Soc. London* (1811) contains, at a rough count, 7 mineralogical articles, no articles on geomorphology, palaeontology or the theory of the Earth, and 10 articles which are primarily regional stratigraphy: MacCulloch (1–22); Holland (38–62); Berger (93–184); Aikin (191–212); Berger (249–68); Fitton (269–80); Horner (281–321); MacCulloch (322–3); Parkinson (324–54); Bennett (391–98). A similar analysis of vol. ii reveals 4 mineralogical articles,

5 palaeontological, but 12 mainly stratigraphical contributions. Vol. iii had 4 mineralogical, no palaeontological, and 9 stratigraphical articles.

13. Horner's article on the Malverns (1811a) tested in the field aspects of Hutton's theory of granite and igneous intrusions. Aikin's paper on the Wrekin (1811) examined whether landforms owed more to elevation or depression. Note Aikin's development from the superficial views of his *Journal* (1797) to the sophisticated geological memoir of later years (1811).

## CHAPTER 8

1. Vol. iii of the Theory, published by Geikie in 1899, in particular demonstrates Hutton's impressiveness as a field geologist. Some of the drawings and sections made for Hutton on his field excursions have recently been discovered and it is hoped will be published under the direction of Professor Gordon Craig. For Hutton's eminence as a fieldworker, cf. V. A. Eyles (1970) and Bailey (1967).

2. Hutton, 1788: 226. Of course this was precisely the kind of claim against which his critics reacted. The fourth edition of the *Encyclopaedia Britannica* ('Geology') stated that geology was a science which could be studied with 'but little advantage in the closet' (ix: 550).

3. I take my summary from the 1788 version. Hutton's 1785 *Abstract* is even more markedly teleological, theological and deductive in form. Cf. Dean (1975).

4. My emphasis upon Hume and Smith as the milieu of Hutton is somewhat conjectural. In no way is it meant to minimize the importance of Hutton's known, real and tangible connections with others (e.g. Black).

5. *Investigation*, iii, 258. Contrast for example William Smith, who thought 'the Book of Nature, which is open to every man, is plain enough for him to "read as he runs"' (Cox, 1942–5: 84); or Whitehurst (1778, ii). The parallel is close between Hutton's view and Hume's division of the world into shallow and abstruse thinkers in his 'Essay on commerce', or Smith's tracing of the gradual emergence of a philosophical understanding of phenomena out of common sense in his 'Essay on the history of astronomy' (1795).

6. To such an extent did Hutton employ final causes – 'it must not be alleged', he wrote (1792: 261), 'that natural philosophy is not concerned with final causes' – that de Luc upbraided him for it. Hutton had argued that de Luc's notion of the land successively collapsing would show 'little wisdom', to which de Luc replied, 'We are too short-sighted, Sir, to claim *wisdom* as an argument *a priori*' (1790–1: 226).

7. For the same reasons, Hutton, no less than most eighteenth-century theorists, argued against extinction, and of course had no concept of evolution (1788: 290–1).

8. It is proper to use Hutton's other writings to help illuminate his geology because he mulled over all these topics during a period of many years, and frequently repeats themes throughout the works. Cf.

Playfair (1805: 5–52 and 77–8). Patsy Gerstner has begun to study Hutton's theories of matter and heat (1968), and Heimann and McGuire his epistemology (1971). Davies (1969) is the first historian of geology to see the importance of Hutton's general physical writings for his geo-philosophy.

9. 1831: 157f.; cf. 1831: 46f. I am grateful for conversations on de Luc with Miss C. C. Orr of Girton College, Cambridge, who is writing a Ph.D. dissertation on him.

10. Cf. John Hunter, 'Our mode of reasoning on this subject may be termed retrograde; it is by supposing, from the state of the earth now, what must have taken place formerly' (Owen, 1861, i: 302). Cf. de Luc (1790–1: 226) and Hutton (1795, i: 4, 10, 36, 41, 373). Leroy E. Page (1969) has well shown how, despite the popular stereotype, English diluvialist geology was generally actualist and naturalistic in its methodology.

11. Particularly important in this period of revitalized debate was the founding of the new monthly scientific periodicals, such as the *Phil. mag.* and the *Ann. phil.* Geological reporting and debates became central to these journals between about 1808 and 1820.

12. 'A constant reader' (1815). On the early nineteenth-century popularity of geology see Gillispie (1951: 116) and Davies (1969: 201). 'Provincial societies have been formed, periodical works teem with enlightened geological essays, and the civil engineer understands well the nature and the position of the beds in which he has to conduct his operation' [Henry Warburton], Report of the Council to the Geological Society of London (1814, Greenough MSS).

13. 'New discoveries in the natural world suggest various applications to practical purposes; new political situations, commonly produced and attended by violence, like natural convulsions, disclosing the mineral strata, illustrate the origin and fate of human society' (*Analytical rev.*, 1797: 105).

## CONCLUSION

1. Nixon (1757). Examples of strong continuities of eighteenth- and nineteenth-century study are legion. The palaeontologically crucial Stonesfield vertebrate remains, explored by Buckland, Conybeare, Cuvier and Owen (Rudwick, 1972: 145f.), had already been described by Platt (1758). Buckland made extensive use of Catcott's work in his celebrated exposition of alluvium and diluvium.

2. 'If only the Geologists would let me alone, I could do very well, but those dreadful Hammers! I hear the clink of them at the end of every cadence of the Bible verses.' John Ruskin, quoted in Burrow (1968: 20): a letter of 24 May 1851 to Henry Acland, in *The works of John Ruskin* London, 1903–12, xxxvi: 115.

3. Burrow notes 'bughunting was the Trojan horse of Victorian agnosticism'. One might add fossil collecting (1968: 19).

# Bibliographical Note

This note attempts to spell out briefly some of the historiographical assumptions which pervade this book, and to discuss sources used and interpretations which I have found helpful.

1. I know of no extensive bibliographies listing published work on the history of geology for the period covered in this book. Bassett (1973) provides a useful general survey of recent work in the broad history of geology. Gillispie (1951) has a helpful bibliographical essay, mainly for the nineteenth century. Challinor (1971) admirably lists the major primary sources for the period, but is slimmer on recent historical works. For older primary bibliographies, see Schall (1787) and Agassiz and Strickland (1843–54). For a checklist of bibliographies of the history of natural history, see Bridson and Harvey (1971). The present book is based on my Cambridge Ph.D thesis (1974), which deals in greater detail and with more comprehensive documentation and bibliography with many of the issues covered.

2. The received, largely Whiggish, account of the history of British geology is to be found in Lyell (1830), Geikie (1897; rep. 1962) and Adams (1938). It is still part of the mental fabric of geologists, and continues to be digested into general histories of science. Zittel (1901) and Beringer (1954) offer much the same picture from a continental point of view.

A revisionist historiography, paying more attention to the broader intellectual and social context and more concerned to understand the scientific endeavours of the past on their own terms, rather than through presentist eyes, has been developed by Hooykaas (1963), Cannon (1960a; 1960b; 1961; 1964a; 1964b), Davies (1969), the contributors to Schneer (1969) and Rudwick (1972). These may be seen as building upon the extremely important biographical and bibliographical foundations laid for any history of British geology by

the indispensable and indefatigable researches of Dr John Challinor
and Dr V. A. and Mrs J. M. Eyles.

3.   Amongst historians and sociologists of science, I have found the
attempts of Kuhn (1963), Ravetz (1971) and Barnes (1974) to provide
a descriptive sociology/social history of science repeatedly germane
and fruitful. S. E. Toulmin (1972) has helped me to address the
question of how and why ideas change, and is a useful foil to
Foucault (1970). Foucault's insistence on the *depth* of epistemological
transformations underpins my argument for the distinctive newness of
geology. Whereas Foucault, however, seems to think such transformations are necessarily mysterious, I have tried in a more conventional
way to plot them (cf. Rudwick, 1975). Young's Marxist position
(1973), that science as part of a society must be seen as constitutive of
its economy, its ideology and its class relations, I take programmatically as a starting-point for my work. I cannot judge how far
Marxist historians would either accept my argument or espouse my
conclusions. Certainly my emphasis on the 'making' of the science
reflects my own intellectual debts to the *Theses on Feuerbach*, as well
as to another, more recent, Marxist 'making' (Thompson, 1968).

4.   We are fortunate to have several extremely stimulating approaches
in intellectual history to the broad problems of changing concepts of
Nature as the object of science. Humboldt (1845–62) and Glacken
(1967) have proved immensely illuminating on transformations in
perceptions of the Earth as habitat. Lovejoy's writings (e.g. 1936;
1948) have focused attention on human classifications of Nature.
Greene (1959) has pinpointed clashes between static and dynamic
conceptions of Creation. Foucault (1970) has helped to conceptualize
a shift from natural history to geology. Heimann and McGuire
(1971), Schofield (1970) and Thackray (1970a) have all showed the
richness of eighteenth-century conceptions of Nature in respect of
God and providence. This book is very deeply indebted to the
problems and approaches these have suggested.
    The social history of science, however – and especially the social
history of the natural history sciences, and of eighteenth-century
English science – is altogether more thinly covered. Here, almost the
only work of importance is D. E. Allen (1976). Although it appeared
too late to influence the course of my own work, I should be very
happy if my book were taken as broadly fitting within the social-
historical approach to natural knowledge, and the broad framework
of interpretation of natural history in England which Mr Allen sets
out.

# Bibliography

This bibliography lists most of the titles referred to in the text. For the sake of brevity, the subtitles of most books have been omitted, and the titles of certain seventeenth- and eighteenth-century books and *Philosophical transactions* articles have been contracted. Journal titles are abbreviated as follows:

| | |
|---|---|
| *Ann. phil.* | *Annals of philosophy* |
| *Ann. sci.* | *Annals of science* |
| *Brit. jnl hist. sci.* | *British journal for the history of science* |
| *Bull. Geol. Soc. America* | *Bulletin of the Geological Society of America* |
| *Edin. rev.* | *Edinburgh review* |
| *Hist. sci.* | *History of science* |
| *Jnl hist. ideas* | *Journal of the history of ideas* |
| *Jnl Soc. Bibliog. Nat. Hist.* | *Journal of the Society for the Bibliography of Natural History* |
| *Notes and records* | *Notes and records of the Royal Society of London* |
| *Phil. mag.* | *Philosophical magazine* |
| *Phil. trans.* | *Philosophical transactions of the Royal Society of London* |
| *Proc. Amer. Phil. Soc.* | *Proceedings of the American Philosophical Society* |
| *Proc. Geol. Asscn* | *Proceedings of the Geological Association* |
| *Proc. Geol. Soc. London* | *Proceedings of the Geological Society of London* |
| *Trans. Geol. Soc. London* | *Transactions of the Geological Society of London* |

*Trans. Roy. Geol. Soc. Cornwall*    *Transactions of the Royal Geological Society of Cornwall*

*Trans. Roy. Irish Acad.*    *Transactions of the Royal Irish Academy*

*Trans. Roy. Soc. Edinburgh*    *Transactions of the Royal Society of Edinburgh*


Academy of Physics. Correspondence Book (National Library of Scotland, MS 755).

Academy of Physics. Extracts from the Minutes. Vol. 1st, 1797, 1798, 1799 (MS in the Royal Society of Edinburgh).

Adams, F. D. 1933a. Earliest use of the term geology. *Bull. Geol. Soc. America.* xliii, 121–3.

Adams, F. D. 1933b. Further note on the use of the term geology. *Bull. Geol. Soc. America.* xliv, 821–6.

Adams, F. D. 1954. *The birth and development of the geological sciences.* New York. 1st ed., London, 1938.

Adamson, I. R. 1976. The foundation and early history of Gresham College, London, 1596–1704. Ph.D. thesis. University of Cambridge.

Addison, J. & R. Steele. 1911. *The Spectator.* 4 vols. Everyman ed. London.

Agassiz, L. & H. E. Strickland. 1843–54. *Bibliographia zoologiae et geologiae.* 4 vols. London.

Agricola, Georgius. 1950. *De re metallica.* Trans. by H. C. and L. H. Hoover. New York. 1st ed., 1556.

Aikin, A. 1797. *Journal of a tour through North Wales and part of Shropshire.* London.

Aikin, A. 1811. Observations on the Wrekin, and on the great coalfield of Shropshire. *Trans. Geol. Soc. London.* i, 191–212.

Akenside, M. 1744. *The pleasures of imagination.* London.

Allan, T. 1813. Remarks on the Transition rocks of Werner. *Phil. mag.* xlii, 15–25.

Allan T. 1814. Answer to Dr Grierson's observations on Transition rocks. *Ann. Phil.* iii, 109–16.

Allen, D. C. 1949. The legend of Noah. *University of Illinois studies in language and literature.* xxxiii.

Allen, D. E. 1976. *The naturalist in Britain.* London.

Allen, M. 1964. *The Tradescants: their plants, garden and museum.* London.

Allen, P. 1949. Scientific studies in the English universities of the seventeenth century. *Jnl hist. ideas.* x, 219–53.

Anderson, J. P. 1881. *The book of British topography.* London.

Anon. 1947. Lusus naturae. *Notes and records.* 42.

Arbuthnot, J. 1697. *An examination of Dr Woodward's account of the Deluge.* London.

Arden, J. 1772. *A short account of a course of natural and experimental philosophy.* Coventry.

Arderon, W. 1746. Extract of a letter from Mr William Arderon, F.R.S., to Mr Henry Baker, F.R.S., containing observations on the precipices or cliffs on the North East sea coast of the County of Norfolk. *Phil. trans.* xliv, 275–84.

Arderon, W. Manuscript remains of William Arderon, Esq., F.R.S. (British Museum, Add. MS 27966).

Arkell, W. J. & S. I. Tomkeieff. 1953. *English rock terms, chiefly as used by miners and quarrymen.* London.

Armitage, A. 1966. *Edmond Halley.* London.

Aubin, R. A. 1934. Grottoes, geology and the Gothic revival. *Studies in philology.* xxxi, 408–16.

Aubin, R. A. 1936. *Topographical poetry in eighteenth century England.* New York.

Aubrey, J. 1685. Memoires of natural remarques in the county of Wiltshire (Manuscript in Royal Society).

Aubrey, J. 1847. *The natural history of Wiltshire.* Ed. by J. Britton. London.


Babbage, C. 1830. *Reflections on the decline of science in England and on some of its causes.* London.

Babington, W. 1795. *A systematic arrangement of minerals.* London.

Babington, W. 1799. *A new system of mineralogy, in the form of a catalogue.* London.

Bacon, F. 1605. *Of the proficience and advancement of learning, divine and humane.* London.

Bagrow, L. 1964. *The history of cartography.* Trans. by D. L. Paisley. London.

Bailey, Sir E. B. 1952. *Geological Survey of Great Britain.* London.

Bailey, Sir E. B. 1967. *James Hutton, the founder of modern geology.* London.

Baker, D. E. 1748. A letter from Mr David Erskin Baker to Martin Folkes Esq., Pr. R. S. containing considerations on two extraordinary Belemnitae. *Phil. trans.* xlv, 598–601.

Baker, H. 1747. A description of a *curious Echinites*; by Mr Henry Baker, F.R.S. *Phil. trans.* xliv, 432–4.

Baker, H. 1749. A letter from Mr Henry Baker to the President, concerning some *Vertebrae of Ammonitae*, or Cornua Ammonis. *Phil. trans.* xlvi, 37–9.

Baker, H. 1753. An account of some uncommon fossil bodies. *Phil. trans.* xlviii, 117–23.

Bakewell, R. 1813. *An introduction to geology*. London.

Bakewell, R. 1815. Observations on the geology of Northumberland and Durham. *Phil. mag.* xlv, 81–96.

Bakewell, R. 1819. *An introduction to mineralogy*. London.

Barba, A. 1674. *The art of metals*. Trans. into English by Edward, Earl of Sandwich. London.

Barnes, S. B. 1974. *Scientific knowledge and sociological theory*. London.

Bartholomew, M. 1973. Lyell and evolution: an account of Lyell's response to the prospect of an evolutionary ancestry for man. *Brit. jnl hist. sci.* vi, 261–303.

Barton, R. 1751. *Lectures in natural philosophy*. Dublin.

Bassett, D. 1973. History of geology. In D. N. Wood (ed.), *Use of earth sciences literature*. London.

Bayne-Powell, R. 1951. *Travellers in eighteenth century England*. London.

Beale, Dr J. 1666. Some promiscuous observations, made in Somersetshire. *Phil. trans.* i, 323.

Beattie, L. M. 1935. *John Arbuthnot, mathematician and scientist*. Cambridge, Mass.

Beaumont, J. 1676. Two letters by Mr John Beaumont. . .concerning rock plants and their growth. *Phil. trans.* xi, 724–42.

Beaumont, J. 1683. A further account of some rock plants growing in the lead mines of the Mendip Hills. *Phil. trans.* xiii, 276–9.

Beaumont, J. 1693. *Considerations on a booke, entituled The theory of the earth, publisht by Dr Burnet*. London.

Beddoes, T. 1791. Observations on the affinity between Basaltes and Granite. *Phil. trans.* lxxxi, 48–70.

Bedini, S. 1965. The evolution of science museums. *Technology and culture*. vi, 1–29.

Beer, Sir G. de. 1952. John Strange, F.R.S., 1732–99. *Notes and records*. ix, 96–108.

Beer, Sir G. de. 1953. *Sir Hans Sloane and the British Museum*. London.

Beer, Sir G. de. 1962. The volcanoes of Auvergne. *Ann. sci.* xviii, 49–61.

Bellers, F. 1712. A description of the several strata of earth, stone, coal, etc, found in a coal pit at the West end of Dudley in Staffordshire. *Phil. trans.* xvii, 541–4.

Ben-David, J. 1971. *The scientist's role in society: a comparative study*. Englewood Cliffs.

Bennett, Hon. H. G. 1811. Sketch of the geology of Madeira. *Trans. Geol. Soc. London.* i, 391–8.

Bentley, R. 1692–3. *The folly and unreasonableness of atheism demonstrated*. Boyle Lectures. London.

Berger, J. F. 1811a. Observations on the physical structure of Devon and Cornwall. *Trans. Geol. Soc. London.* i, 99–184.

Berger, J. F. 1811b. A sketch of the geology of some parts of Hampshire and Dorsetshire. *Trans. Geol. Soc. London.* i, 249–68.

Berger, J. F. 1814. Mineralogical account of the Isle of Man. *Trans. Geol. Soc. London.* ii, 29–65.

Bergman, T. 1766. *Physisk Beskrifning öfver Jordlotet*. Upsala.

Bergman, T. 1783. *Outlines of mineralogy*. Trans. by W. Withering. London.

Beringer, C. C. 1954. *Geschichte der Geologie und des geologisches Weltbildes*. Stuttgart.

Berman, M. 1972. The early years of the Royal Institution, 1799–1810: a re-evaluation. *Science studies*. ii, 205–40.

Berry, H. F. 1915. *A history of the Royal Dublin Society*. London.

Best, M. R. & R. H. Brightman (eds.) 1973. *The book of secrets of Albertus Magnus*. Oxford.

Birch, T. 1756–7. *A history of the Royal Society of London*. 4 vols. London.

Biswas, A. K. 1972. *History of hydrology*. Amsterdam.

Black, J. Papers in the Edinburgh University Library: Gen 813/111/142, f. 71–2 (1788), f. 200 (1791).

Blacker, C. & M. Loewe (eds.) 1975. *Ancient cosmologies*. London.

Bloor, D. 1976. *Scientific knowledge and social imagery*. London.

Boate, G. 1652. *Irelands naturall history*. London.

Borelli, J. A. 1671. Historia & meteorologia incendii Aetnaei. *Phil. trans.* v, 2264–8.

Borlase, W. 1753. An account of the great alterations which the Islands of Sylley have undergone since the time of the Ancients. *Phil. trans.* xlviii, 55–69.

Borlase, W. 1756. *Observations on the ancient and present state of the Islands of Scilly*. Oxford.

Borlase, W. 1758. *The natural history of Cornwall*. Oxford.

Bowles, W. 1783. *Introduzione alla storia naturale e alla geografia fisica di Spagna*. Parma.

Boyle, R. 1672. *An essay about the origine and virtues of gems*. London.

Boyle, R. 1774. *Works – with a life of the author by T. Birch*. London.

Boyne, W. 1869. *The Yorkshire library*. London.

Brande, W. T. 1816. *A descriptive catalogue of the British specimens deposited in the geological collection of the Royal Institution*. London.

Brander, G. 1754. A dissertation on the Belemnites. *Phil. trans.* xlviii, 803–10.

Brander, G. 1766. *Fossilia Hantoniensia.* London.

Bridson, G. D. R. & A. Harvey 1971. Checklist of natural history bibliographies and bibliographical scholarship, 1966–70. *Jnl Soc. Bibliog. Nat. Hist.* v, 428–67; vi, 263–92.

Brooks, E. St John. 1954. *Sir Hans Sloane: the great collector and his circle.* London.

Brown, F. K. 1961. *Fathers of the Victorians: the age of Wilberforce.* Cambridge.

Brown, Harcourt. 1934. *Scientific organizations in seventeenth century France, 1620–1680.* Baltimore.

Browne, E. 1669. An extract of a letter. . .concerning damps in the mines of Hungary and their effects. *Phil. trans.* iv, 965–7.

Browne, E. 1670a. Concerning the mines, minerals, baths, etc. of Hungary, Transylvania, Austria. *Phil. trans.* v, 1189–90.

Browne, E. 1670b. An accompt concerning the baths of Austria and Hungary. *Phil. trans.* v, 1044–51.

Browne, E. 1673. *A brief account of some travels in Hungaria, Servia, Bulgaria.* London.

Browne, E. 1677. *An account of several travels through a great part of Germany, in four journeys.* London.

Browne, Sir T. 1928–1931. *Works.* Ed. by G. Keynes. 6 vols. London.

Brydone, P. 1773. *A tour through Sicily and Malta.* London.

Buchdahl, G. 1969. *Metaphysics and the philosophy of science.* Oxford.

Buckland, W. 1820. *Vindiciae geologicae: or the connexion of geology with religion explained.* Oxford.

Buckland, W. 1821. Description of the quartz rock of the Lickey Hill in Worcestershire. *Trans. Geol. Soc. London.* v, 506–44.

Buckland, W. 1823. *Reliquiae diluvianae.* London.

Buffon, G. L. L. Comte de. 1749–1803. *Histoire naturelle.* 44 vols. Trans. by W. Smellie, 9 vols., Edinburgh, 1785.

Bulkeley, Sir R. 1693. Part of a letter from Sir R. Bulkeley to Dr Lister, concerning the Giants Causeway. *Phil. trans.* xvii, 708–10.

Burchfield, J. D. 1975. *Lord Kelvin and the age of the Earth.* London.

Burke, J. G. 1966. *Origins of the science of crystals.* Berkeley.

Burnet, T. 1684. *The theory of the Earth.* London.

Burnet, T. 1690. *An answer to the late exceptions made by Mr Erasmus Warren against the Theory of the Earth.* London.

Burnet, T. 1691. *A short consideration of Mr Erasmus Warren's defence of his Exceptions against the Theory of the Earth, in a letter to a friend.* London.

Burnet, T. 1699. *Reflections upon the Theory of the earth, occasion'd by a late examination of it.* London.

Burrow, J. W. 1968. *The origin of species by Charles Darwin,* edited with an introduction by *J. W. Burrow.* London.

Burt, R. 1969. *Cornish mining.* Newton Abbot.

Burtt, E. A. 1925. *The metaphysical foundations of modern physical science.* London.

Bush, R. 1974. The development of geological mapping in Britain, 1795–1825. Ph.D thesis. University of London.

Butler, H. K. 1968. The study of fossils in the last half of the seventeenth century. Ph.D. thesis. University of Oklahoma.

Butts, R. E. & J. W. Davis (eds.) 1970. *The methodological heritage of Newton.* Oxford.

Bynum, W. F. 1974. Time's noblest offspring: the problem of man in the British natural historical sciences 1800–1863. Ph.D. thesis. University of Cambridge.

Bynum, W. F. 1975. The Great Chain of Being after forty years. *Hist. sci.* xiii, 1–8.

Bynum, W. F. 1976. The blind man and the elephant: toward a history of pre-history. Unpublished paper delivered to the Conference on New Perspectives in the History of the Life Sciences, Cambridge, England, March 1976.

Calinger, R. S. 1968. Frederick the Great and the Berlin Academy of Sciences, 1740–1766. *Ann. sci.* xxiv, 239–49.

Camden, W. 1586. *Britannia.* London. English trans. by P. Holland, London, 1610.

Cameron, H. C. 1956. *Sir Joseph Banks: the autocrat of the philosophers. 1744–1820.* London.

Cannon, W. F. 1960a. The Uniformitarian–Catastrophist debate. *Isis.* li, 38–55.

Cannon, W. F. 1960b. The problem of miracles in the 1830s. *Victorian studies.* iv, 5–32.

Cannon, W. F. 1961. The impact of Uniformitarianism. *Proc. Amer. Phil. Soc.* cv, 301–14.

Cannon, W. F. 1964a. Scientists and Broad Churchmen: an early Victorian intellectual network. *Journal of British studies.* iv, 65–88.

Cannon, W. F. 1964b. The normative role of science in early Victorian thought. *Jnl hist. ideas.* xxv, 487–502.

Cantor, G. 1974. The Academy of Physics at Edinburgh, 1797–1800. *Social studies of science.* v, 109–34.

Cardwell, D. S. L. 1972. *The organisation of science in England.* London.

Carew, R. 1602. *The survey of Cornwall*. London.

Carozzi, A. V. 1969. Rudolf Erich Raspe and the basalt controversy. *Studies in Romanticism*. viii, 235–50.

Carr, E. H. 1961. *What is history?* Harmondsworth.

Carr, J. 1974. The biological works of Martin Lister 1639–1712. Ph.D. thesis. University of Leeds.

Carswell, J. P. 1950. *The prospector: being the life and times of R. E. Raspe, 1737–94*. London.

Cat, N. le. 1750. An account of several systems: particularly that of the ingenious Mr Le Cat, with regard to the formation of mountains. *Monthly review*, iii, 375–93; 444–59.

Catcott, A. Manuscript of Tours (Bristol Central Library, MS B 6495).

Catcott, A. 1761. *A treatise on the Deluge*. London.

Catcott, A. S. 1822. *The antient principles of the true and sacred philosophy*. Trans. by Alexander Maxwell. London. First ed., 1738.

Challinor, J. 1953. The early progress of British geology. I. From Leland to Woodward, 1538–1728. *Ann. sci.* ix, 124–53.

Challinor, J. 1954. The early progress of British geology. II. From Strachey to Michell, 1719–1788. *Ann. sci.* x, 1–19.

Challinor, J. 1971. *The history of British geology: a bibliographical study*. Newton Abbot.

Chalmers, G. K. 1936. Sir Thomas Browne: true scientist. *Osiris*. ii, 28–79.

Chambers, E. 1741–3. *Cyclopaedia*. 2 vols. 5th ed. London.

*Chartham News*. 1701. Chartham news, or a brief relation of some strange bones there lately digged up, in some grounds of Mr John Sommers in Canterbury. *Phil. trans.* xxii, 882–93.

Childe-Pemberton, W. S. 1924. *The earl bishop: the life of Frederick Hervey, Bishop of Derry and Earl of Bristol*. 2 vols. London.

Childrey, J. 1661. *Britannia Baconica*. London.

Chitnis, A. C. 1970. The University of Edinburgh's Natural History Museum and the Huttonian–Wernerian debate. *Ann. sci.* xxvi, 85–94.

Christie, J. R. R. 1974. The origins and development of the Scottish scientific community, 1680–1760. *Hist. sci.* xii, 122–41.

Clarke, J. W. & T. M. Hughes. 1890. *The life and letters of A. Sedgwick*. 2 vols. Cambridge.

Clark, K. M. 1949. *Landscape into art*. London.

Clarke, Sir E. 1898. The Board of Agriculture. *Jnl of the Royal Agricultural Society*. 3rd ser. ix, 1–41.

Clarke, E. D. 1793. *A tour through the South of England, Wales and part of Ireland, made during the Summer of 1791*. London.

Clarke, E. D. 1810–23. *Travels in various countries of Europe, Asia and Africa.* London.

Clayton, R., Bishop of Clogher. 1752. *A vindication of the histories of the Old and New Testament.* Dublin.

Clerk, Sir J. 1892. *Memorials of the life of Sir John Clerk of Penicuik.* Ed. by John M. Gray. Scottish History Society Publication. Edinburgh.

Clifford, J. L. (ed.) 1968. *Man versus society in eighteenth century Britain.* Cambridge.

Clive, J. L. 1957. *Scotch reviewers: the Edinburgh review, 1802–15.* London.

Clow, A. & N. L. 1952. *The chemical revolution.* London.

Cochrane, A., Earl of Dundonald. 1793. *Description of the Estate and Abbey of Culross.* Edinburgh.

Cockburn, H. 1856. *Memorials of his time.* Edinburgh.

Colepresse, S. 1667. Extract of a letter written by Mr Sam. Colepress. *Phil. trans.* ii, 500–1.

Colepresse, S. 1671. An account of some mineral observations touching the mines of Cornwall. *Phil. trans.* vi, 2096–113.

Colie, R. L. 1957. *Light and enlightenment. A study of the Cambridge Platonists and the Dutch Arminians.* Cambridge.

Collier, K. B. 1934. *Cosmogonies of our fathers.* New York.

Collingwood, W. G. (ed.) 1912. *Elizabethan Keswick.* Kendal.

Constant Reader. 1815. An earnest recommendation to curious ladies and gentlemen residing or visiting in the country, to examine the quarries, cliffs, steep banks etc., and collect and preserve fossil shells. *Phil. mag.* xlv, 274–80.

Conybeare, W. D. 1811. Letter to G. B. Greenough, 18 June (Cambridge University Library, Greenough MSS).

Conybeare, W. D. & H. T. de la Beche. 1821. Notice of the discovery of a new fossil animal. *Trans. Geol. Soc. London.* v, 559–94.

Conybeare, W. D. & W. Phillips. 1822. *Outlines of the geology of England and Wales.* Part i. London.

Cox, E. G. 1939–49. *A reference guide to the literature of travel (including tours, descriptions, towns, histories and antiquities).* 3 vols. Seattle.

Cox, L. R. 1942–45. New light on William Smith and his work. *Proc. of the Yorkshire Geological Society.* xxv, 1–99.

Cragg, G. R. 1950. *From Puritanism to the Age of Reason.* Cambridge.

Cragg, G. R. 1964. *Reason and authority in the eighteenth century.* Cambridge.

Crane, R. S. 1934. Anglican apologetics and the idea of progress. *Modern philology.* xxxi, 273–306; 349–82.

Croft, H. 1685. *Some animadversions upon a book entituled the Theory of the earth.* London.

Cronstedt, A. F. 1788. *An essay towards a system of mineralogy.* Trans. by E. M. da Costa, 1777; improved by M. Magellan, London. 1st ed., 1772.

Crosland, M. (ed.) 1975. *The emergence of science in Western Europe.* London.

Cudworth, R. 1678. *The true intellectual system of the universe.* London.

Cullmann, O. 1951. *Christ and time: the primitive Christian conception of time and history.* Trans. by F. V. Filson. London.

Dance, S. P. 1962–8. The authorship of the Portland catalogue (1786). *Jnl Soc. Bibliog. Nat. Hist.* iv, 30–4.

Dance, S. P. 1966. *Shell collecting: an illustrated history.* London.

Daniel, G. E. 1961. *The idea of prehistory.* London.

Daniell, W. V. & F. J. Nield. 1909. *Manual of British topography: a catalogue of county and local histories, pamphlets, views, drawings, maps etc.* London.

Darwin, E. 1801. *The botanic garden.* London.

Darwin, E. 1803. *The temple of Nature, or the origin of society: a poem with philosophical notes.* London.

Davie, G. E. 1961. *The democratic intellect.* Edinburgh.

Davies, G. L. 1964. Robert Hooke and his conception of earth history. *Proc. Geol. Asscn.* lxxv, 493–8.

Davies, G. L. 1966a. The concept of denudation in seventeenth century England. *Jnl hist. ideas.* xxvii, 278–84.

Davies, G. L. 1966b. The eighteenth century denudation dilemma and the Huttonian theory of the earth. *Ann. sci.* xxii, 129–38.

Davies, G. L. 1966c. Early British geomorphology, 1578–1705. *Geographical jnl.* cxxxii, 252–62.

Davies, G. L. 1967. George Hoggart Toulmin and the Huttonian *Theory of the earth. Bull. Geol. Soc. America.* lxxviii, 121–3.

Davies, G. L. 1969. *The earth in decay: a history of British geomorphology.* London.

Dawson, W. R. 1958. *The Banks letters; a calendar.* London.

Dean, D. R. 1973. James Hutton and his public, 1785–1802. *Ann. sci.* xxx, 89–105.

Dean, D. R. 1975. James Hutton on religion and geology: the unpublished preface to his *Theory of the earth,* 1788. *Ann. sci.* xxxii, 187–93.

Debus, A. G. 1961. Gabriel Plattes and his chemical theory of the formation of the earth's crust. *Ambix.* ix, 162–5.

Debus, A. G. 1965. *The English Paracelsians.* London.

Debus, A. G. 1966. Renaissance chemistry and the work of Robert Fludd. In A. G. Debus & R. P. Multhauf (eds.), *Alchemy and chemistry in the seventeenth century*. Los Angeles.

Debus, A. G. 1967. Fire analysis and the elements in the sixteenth and seventeenth centuries. *Ann. sci.* xxiii, 127–47.

Debus, A. G. 1969. Edward Jorden and the fermentation of the metals. An iatrochemical study of terrestrial phenomena. In C. J. Schneer (ed.), *Toward a history of geology*. Cambridge, Mass.

Debus, A. G. 1974. The chemical philosophers: chemical medicine from Paracelsus to Van Helmont. *Hist. sci.* xii, 235–59.

Debus, A. G. 1977. *The chemical philosophy*. New York.

Defoe, D. 1724–7. *A tour thro' the whole island of Great Britain*. 3 vols. London.

Dixon, J. 1801. *The literary life of William Brownrigg, M.D.* London.

Dobrée, B. 1954. *The broken cistern*. London.

Donald, M. B. 1955. *Elizabethan copper: the history of the Company of Mines Royal*. London.

Donald, M. B. 1961. *Elizabethan monopolies: the history of the Company of Mineral and Battery Works from 1565 to 1604*. Edinburgh.

Douglas, D. C. 1939. *English scholars, 1660–1730*. London.

Douglas, Rev. J. 1785. *A dissertation on the antiquity of the earth*. London.

Douglas, M. 1970. *Natural symbols: explorations in cosmology*. London.

Duckham, B. F. 1970. *A history of the Scottish coal industry. Vol. i. 1700–1815*. Newton Abbot.

Duncan, E. H. 1954. The natural history of metals and minerals in the universe of Milton's *Paradise lost*. *Osiris*. xi, 386–421.

Edwards, E. 1870. *Lives of the founders of the British Museum*. 2 vols. London.

Edwards, G. 1776. *Elements of fossilogy*. London.

Ellis, A. 1956. *The penny universities*. London.

Emery, F. V. 1971. *Edward Lhuyd, 1660–1709*. Caerdydd.

Engineer. 1815. On the damps in mines. *Phil. mag.* xlv, 116–18.

Ercker, L. 1683–6. *Treatise on ores and assaying*. Trans. by Sir J. Pettus in his *Fleta minor*. London.

'Espinasse, M. 1958. The decline and fall of Restoration science. *Past and present*. xiv, 71–89.

Evelyn, J. 1665. An advertisement of a way of making more lively counterfeits of Nature in wax than are extant in painting. And of a new kind of maps in a low relievo. Both practised in France. *Phil. trans.* i, 99–100.

Eyles, J. M. 1961. Carl Ludwig Giesecke, 1761–1833. *Nature*, cxc, 213.

Eyles, J. M. 1969. William Smith: some aspects of his life and work. In C. J. Schneer (ed.), *Toward a history of geology*. Cambridge, Mass. 142–58.

Eyles, V. A. 1958. The influence of Nicolaus Steno on the development of geological science in Britain. In G. Scherz (ed.), *Nicolaus Steno and his Indice*. Copenhagen. 167–88.

Eyles, V. A. 1961. Sir James Hall, Bt, 1761–1832. *Endeavour.* xx, 210–16.

Eyles, V. A. 1963. The evolution of a chemist. *Ann. sci.* xix, 153–82.

Eyles, V. A. 1969. The extent of geological knowledge in the eighteenth century and the methods by which it was diffused. In C. J. Schneer (ed.), *Toward a history of geology*. Cambridge, Mass. 159–83.

Eyles, V. A. 1970. Ed. of *James Hutton's System of the earth 1785* etc. New York.

Eyles, V. A. 1971. John Woodward, F.R.S., F.R.C.P., M.D., (1665–1728), a bio-bibliographical account of his life and work. *Jnl Soc. Bibliog. Nat. Hist.* v, 399–427.

Eyles, V. A. & J. M. 1951. Some geological correspondence of James Hutton. *Ann. sci.* vii, 316–39.


Farey, J. 1807. On the dislocations of the strata of the earth. *Phil. mag.* xxviii, 120.

Farey, J. 1810. Copy of a list of the principal British strata, by the late Rev. John Michell. *Phil. mag.* xxxvi, 102–4.

Farey, J. 1811–15. *General view of the agriculture and minerals of Derbyshire.* 3 vols. London.

Farey, J. 1811. On Dr Walker's opinion respecting the general Deluge. *Phil. mag.* xxxviii, 35–9.

Farey, J. 1812. Letter to G. B. Greenough, 7 Jan. (Cambridge University Library, Greenough MSS).

Farey, J. 1813. Cursory geological observations made in Shropshire, Wales, Lancashire, Scotland, Durham, Yorkshire, North Riding, and Derbyshire. *Phil. mag.* xlii, 53–9.

Farey, J. 1814. Notes and observations on the remaining part of the sixth and part of the seventh chapters of Mr Robert Bakewell's *Introduction to geology. Phil. mag.* xliii, 27–34; 119–27; 182–90; 252–61; 325–41.

Farey, J. 1815a. Short notices of geological observations made. . .in the South of Yorkshire. *Phil. mag.* xlv, 161–77.

Farey, J. 1815b. Observations on the priority of Mr Smith's investigations of the strata of England. *Phil. mag.* xlv, 333–44.

Farey, J. 1818. Mr W. Smith's discoveries in geology. *Ann. phil.* xi, 359–64.

Farey, J. 1819. Free remarks on the geological work of Mr Greenough. *Phil. mag.* liv, 127–32.

Farey, J. 1820. Free remarks on Mr Greenough's geological map, lately published under the direction of the Geological Society of London. *Phil. mag.* lv, 379–83.

Farrar, D. M. 1971. The Royal Hungarian Mining Academy, Schemnitz. Some aspects of technical education in the eighteenth century. M.Sc. thesis. University of Manchester.

Farrell, M. 1973. The life and work of William Whiston. Ph.D. thesis. University of Manchester.

Figlio, K. 1975. Theories of perception and the physiology of mind in the late eighteenth century. *Hist. sci.* xii, 177–212.

Fitton, W. 1811. Notice respecting the geological structure of the vicinity of Dublin. *Trans. Geol. Soc. London.* i, 269–80.

[Fitton, W.] 1817a. Review of *Trans. Geol. Soc. London,* ii. *Edin. rev.* xxviii, 174–92.

[Fitton, W.] 1817b. Review of *Trans. Geol. Soc. London,* iii. *Edin. rev.* xxix, 70–94.

[Fitton, W.] 1818. Review of *A delineation of the strata of England and Wales. Edin. rev.* xxix, 310–37.

[Fitton, W.] 1823. Review of *Reliquiae diluvianae. Edin. rev.* xxxix, 196–234.

Foley, S. 1694. An account of the Giants Causeway in the North of Ireland. *Phil. trans.* xviii, 170–2.

Ford, T. D. 1960. White Watson (1760–1835) and his geological sections. *Proc. Geol. Assocn.* lxxi, 349–63.

Forster, J. R. 1768. *An introduction to mineralogy.* London.

Forster, W. 1809. *Treatise of the strata.* Newcastle. 2nd ed., 1821.

Fothergill, B. 1969. *Sir William Hamilton: envoy extraordinary.* London.

Fothergill, B. 1974. *The mitred earl: an eighteenth century eccentric.* London.

Foucault, M. 1970. *The order of things.* London. English trans. from the French, Paris, 1966.

Frängsmyr, T. 1974. Swedish science in the eighteenth century. *Hist. sci.* xii, 29–42.

Frank, R. G. (Jr) 1973. Science, medicine and the universities of early modern England: background and sources. *Hist. sci.* xi, 194–216; 239–69.

Fuller, J. G. C. M. 1969. The industrial basis of stratigraphy: John Strachey 1671–1743 and William Smith, 1769–1839. *The American Association of Petroleum Geologists' Bulletin.* liii, 2256–2273.

Fussell, G. E. 1935. *The exploration of England: a select bibliography of travel and topography 1570–1815.* London.

Garfinkle, N. 1955. Science and religion in England 1790–1800: The critical response to the work of Erasmus Darwin. *Jnl hist. ideas.* xvi, 376–88.

Gay, P. J. 1967–70. *The Enlightenment: an interpretation.* 2 vols. London.

Geikie, Sir A. 1871. *The Scottish school of geology.* Edinburgh.

Geikie, Sir A. (ed.) 1907. *Faujas de St Fond: A journey through England and Scotland to the Hebrides in 1784.* 2 vols. Glasgow.

Geikie, Sir A. 1918. *Memoir of John Michell.* Cambridge.

Geikie, Sir A. 1962. *The founders of geology.* New York. 1st ed., London, 1897.

*Geological inquiries.* 1808. London. Reprinted in 1817, *Phil. mag.* xlix, 421–9.

Gerstner, P. 1968. James Hutton's *Theory of the earth* and his theory of matter. *Isis.* lxix, 26–31.

Gerstner, P. 1971. The reaction to James Hutton's use of heat as a geological agent. *Brit. jnl hist. sci.* v, 353–62.

Gibbs, F. W. 1960. Itinerant lecturers in natural philosophy. *Ambix.* vii, 111–17.

Gilks, M. 1740. A letter from Mr Moreton Gilks. F.R.S. . . . giving some account of the petrifications near Matlock Baths in Derbyshire. *Phil. trans.* xli, 353–6.

Gillispie, C. C. 1959. *Genesis and geology.* Cambridge, Mass. 1st ed., 1951.

Gillispie, C. C. 1960. *The edge of objectivity.* Princeton.

Glacken, C. J. 1967. *Traces on the Rhodian shore.* Berkeley.

Glanvill, J. 1667. Answers to some of the inquiries formerly publish'd concerning mines. *Phil. trans.* ii, 525–7.

Glanvill, J. 1668. Additional answers to the Queries of Mines. *Phil. trans.* iii, 767–71.

Glass, H. B., O. Temkin & W. L. Straus (eds.) 1959. *Forerunners of Darwin.* Baltimore.

Goldsmith, O. 1774. *An history of the earth and animated nature.* 8 vols. London.

Goltz, D. 1972. *Studien zur Geschichte der Mineralnamen in Pharmazie, Chemie, und Medizin von den Anfängen bis auf Paracelsus.* Wiesbaden.

Gortani, M. 1963. Italian pioneers in geology and mineralogy. *Jnl of world history*. vii, 503–19.

Gough, J. W. 1930. *The mines of Mendip*. Oxford.

Gough, J. W. 1932. *The superlative prodigall: a life of T. Bushell*. Bristol.

Graydon, Rev. G. 1791. On the fish enclosed in stone of Monte Bolca. *Trans. Roy. Irish Acad*. v, 281–317.

Greene, J. C. 1959. *The death of Adam*. Iowa.

Greenough, G. B. 1819a. *A critical examination of the first principles of geology*. London.

Greenough, G. B. 1819b. *A geological map of England and Wales*. London.

Greenough, G. B. MSS (Cambridge University Library).

Grew, N. 1681. *Musaeum Regalis Societatis*. London.

Grew, N. 1701. *Cosmologia sacra*. London.

Guerlac, H. & M. C. Jacob. 1969. Bentley, Newton and Providence (The Boyle Lectures once more). *Jnl hist. ideas*. xxx, 307–18.

Guntau, M. 1967. Der Aktualismus bei A. G. Werner. *Bergakademie*. v, 294–7.

Guntau, M. 1968. Die Bedeutung von Abraham Gottlob Werner für die Mineralogie und die Geologie. *Geologie*. xvii, 1096–115.

Guntau, M. 1969. Die Entwicklung der Vorstellung von der Mineralogie in der Wissenschaftsgeschichte. *Geologie*. xviii, 526–37.

Guntau, M. 1971. Kriterien für die Herausbildung der Lagerstättenlehre als Wissenschaft im 19. Jahrhundert. *Geologie*. xx, 348–61.

Guntau, M. 1974. Die geologische Wissenschaften an der Bergakademie Freiberg in der Periode der industriellen Revolution. *NTM-Schriftenr. Gesch., Naturwiss., Technik. Med*. xi, 16–23.

Gunther, R. W. T. 1925a (vol. iii); 1925b (iv); 1939 (xii); 1945 (xiv). *Early science in Oxford*. Oxford.

Gunther, R. W. T. (ed.) 1928. *Further correspondence of John Ray*. London.

Gunther, R. W. T. 1937. *Early science in Cambridge*. Oxford.

Haber, F. C. 1959. *The age of the world: Moses to Darwin*. Baltimore.

Hagner, A. H. 1963. Philosophical aspects of the geological sciences. In C. C. Albritton (ed.), *The fabric of geology*. Stanford. 233–41.

Hahn, R. 1971. *The anatomy of a scientific institution: the Paris Academy of Sciences 1666–1803*. Berkeley.

Hailstone, J. 1792. *A plan of a course of lectures on mineralogy*. Cambridge.

Hale, M. 1677. *The primitive origination of mankind*. London.

Hall, A. R. & M. B. (eds.) 1965–. *The correspondence of Henry Oldenburg.* 10 vols. to date. Madison.

Hall, Sir J. 1794. Observations on the formation of granite. *Trans. Roy. Soc. Edinburgh.* iii, 8–12.

Hall, Sir J. 1798. Curious circumstances upon which the vitreous or the stony character of whinstone and lava respectively depend. *Nicholson's journal.* ii, 285–8.

Hall, Sir J. 1800. Experiments on whinstone and lava. *Nicholson's journal.* iv, 8–18; 56–65.

Hall, Sir J. 1805. Experiments on whinstone and lava. *Trans. Roy. Soc. Edinburgh.* v, 43–75.

Hall, Sir J. Tours abroad, 9 May 1783 to 14 April 1784 (Manuscript in National Library of Scotland, MS 6324).

Hall, Sir J. Journal, 1791 (National Library of Scotland, MS 6329).

Hall, Sir J. 1812. Letter to Dr Marcet, 7 Feb. (National Library of Scotland, 3818, f. 49).

Hall, Sir J. 1813. Letter to Dr Marcet, 15 Sept. (National Library of Scotland, 3813, f. 51).

Hall, M. B. 1970. *Nature and nature's laws.* London.

Hall, M. B. 1975. Science in the early Royal Society. In M. Crosland (ed.), *The emergence of science in Western Europe.* London. 57–77.

Halley, E. 1691. An hypothesis of the internal structure of the earth. *Phil. trans.* xvii, 563–78.

Halley, E. 1715. A short account of the cause of the saltness of the ocean. *Phil. trans.* xxix, 296–300.

Hamilton, Sir W. 1767. Two letters. . .containing an account of the late eruption of Mount Vesuvius. *Phil. trans.* lvii, 192–200.

Hamilton, Sir W. 1768. An account of the eruption of Mount Vesuvius in 1767. *Phil. trans.* lviii, 1–12.

Hamilton, Sir W. 1769. A letter containing some farther particulars on Mount Vesuvius. *Phil. trans.* lix, 18–22.

Hamilton, Sir W. 1770. An account of a journey to Mount Etna. *Phil. trans.* lx, 1–19.

Hamilton, Sir W. 1771. Remarks upon the nature of the soil of Naples and its neighbourhood. *Phil. trans.* lxi, 1–43.

Hamilton, Sir W. 1772. *Observations on Mount Vesuvius, Mount Etna and other volcanoes.* London.

Hamilton, Sir W. 1776–9. *Campi Phlegraei. Observations on the volcanoes of the Two Sicilies.* Naples.

Hamilton, Sir W. 1778. A letter. . .giving an account of certain traces of volcanos on the banks of the Rhine. *Phil. trans.* lxviii, 1–7.

Hamilton, Sir W. 1780. An account of an eruption of Mount Vesuvius. *Phil. trans.* lxx, 42–84.

Hamilton, Sir W. 1783. An account of the earthquakes which happened in Italy. *Phil. trans.* lxxiii, 169–208.

Hamilton, Sir W. 1786. Some particulars of the present state of Mount Vesuvius. *Phil. trans.* lxxvi, 365–80.

Hamilton, Sir W. 1795. An account of the late eruption of Mount Vesuvius. 73–113.

Hans, N. A. 1951. *New trends in education in the eighteenth century.* London.

Harris, J. 1697. *Remarks on some late papers relating to the Universal Deluge and to the natural history of the earth.* London.

Harris, J. 1708–10. *Lexicon technicum: or a universal English dictionary of arts and sciences.* 2 vols. 2nd ed. London.

Harris, V. I. 1966. *All coherence gone.* London.

Harvey, R. A. 1971. The life of Henry Cavendish, 1731–1810. Ph.D. thesis. University of Sheffield.

Hatchett, C. 1967. *The Hatchett diary.* Ed. by A. Raistrick. Truro.

Hauksbee, F. 1712. A description of the several strata of earth, stone, coal, etc. found in a coal pit at the West End of Dudley in Staffordshire. *Phil. trans.* xxvii, 541–4.

Hawkins, J. 1818. Queries proposed to captains of mines and other persons connected with the practical part of mining. *Trans. Roy. Geol. Soc. Cornwall.* i, 242–9.

Hawkins, J. MSS in the Library of the Royal Geological Society of Cornwall.

Hazard, P. G. M. C. 1964. *The European mind, 1680–1715.* Trans. by J. L. May. Harmondsworth.

Hedberg, H. D. 1969. The influence of Torbern Bergman (1735–1784) on stratigraphy: a resumé. In C. J. Schneer (ed.), *Toward a history of geology.* Cambridge, Mass. 186–91.

Heimann, P. M. 1973. 'Nature is a perpetual worker.' Newton's aether and eighteenth century natural philosophy. *Ambix.* xx, 1–25.

Heimann, P. M. & J. E. McGuire. 1971. Newtonian forces and Lockean powers. Concepts of matter in eighteenth century thought. *Historical studies in the physical sciences.* iii, 233–306.

Heton, T. 1707. *Some account of mines and the advantages of them to this kingdom.* London.

Hibbert, C. 1969. *The Grand Tour.* London.

Hill, Sir J. 1771. *Fossils arranged according to their obvious characters.* London.

Hirst, D. 1964. *Hidden riches: traditional symbolism from the Renaissance to Blake.* London.

Hobbs, W. 1715. The earth generated and anatomized. Weymouth.

Manuscript in possession of the British Museum, Natural History. Text, ed. by R. S. Porter, to be published 1977 in the *Bulletin of the British Museum Natural History (Historical series).*

Hodgen, M. T. 1964. *Early anthropology in the sixteenth and seventeenth centuries.* Philadelphia.

Hoeniger, F. D. & F. J. M. 1969. *The growth of natural history in Stuart England from Gerard to the Royal Society.* Charlottesville, Va.

Holland, H. 1811. A sketch of the natural history of the Cheshire rock-salt district. *Trans. Geol. Soc. London.* i, 38–62.

Holloway, B. 1723. An account of the pits for fullers earth in Bedfordshire. *Phil. trans.* xxxii, 419–21.

Holloway, B. MSS (Bodleian Library, MSS Gough, Wales 8).

Holmes, G. & W. A. Speck. 1967. *The divided society.* London.

Holmes, J. H. H. 1816. *A treatise on the coal mines of Northumberland and Durham.* London.

Home, Sir E. 1817. An account of some fossil remains of the rhinoceros. *Phil. trans.* 176–82.

Home, Sir E. 1818. Additional facts respecting the fossil remains of an animal. *Phil. trans.* 24–32.

Home, Sir E. 1819. An account of the fossil skeleton of the Proteosaurus. *Phil. trans.* 209–11.

Hooke, R. 1665. *Micrographia.* London.

Hooke, R. 1705. *The posthumous works of Robert Hooke.* Publish'd by R. Waller. London.

Hooson, W. 1747. *The miners dictionary.* Wrexham.

Hooykaas, R. 1950. The species concept in eighteenth century mineralogy. *Actes du VI\* congrès internationale d'histoire des sciences.* vi, 458–68.

Hooykaas, R. 1963. *Natural law and divine miracle: the principle of uniformity in geology, biology and theology.* Leiden.

Hooykaas, R. 1972. *Religion and the rise of modern science.* Edinburgh.

Hoppen, K. T. 1970. *The common scientist in the seventeenth century.* London.

Horner, L. 1811a. On the mineralogy of the Malvern Hills. *Trans. Geol. Soc. London.* i, 281–321.

Horner, L. 1811b. Letter to G. B. Greenough, 16 Sept. (University Library, Greenough MSS).

Houghton, T. 1681. *Rara avis in terra.* London.

Houghton, W. E. (Jr) 1942. The English virtuoso in the seventeenth century. *Jnl hist. ideas.* iii, 51–73; 190–219.

Howard, P. 1797. *The Scriptural history of the earth and mankind.* London.

Huddesford, W. 1760. 2nd, augmented edition of E. Lhwyd, *Lithophylacii Britannici ichnographia*. Oxford.

Hughes, A. 1951. Science in English encyclopaedias 1704–1875. *Ann. sci.* vii, 340–70; 1952, viii, 323–67; 1953; ix, 233–64.

Humboldt, A. von. 1845–62. *Kosmos*. Stuttgart. English trans. by A. Pritchard, London, 1845–8.

Hume, D. 1779. *Dialogues concerning natural religion*. London.

Hunter, J. 1794. Account of some remarkable caves in the Principality of Bayreuth and of the fossil bones found therein. *Phil. trans.* 402–17.

Hunter, Rev. J. 1830. *The diary of Ralph Thoresby*. 2 vols. London.

Hunter, Rev. J. 1832. *Letters of eminent men, addressed to Ralph Thoresby*. 2 vols. London.

Hunter, M. 1975. *John Aubrey and the realm of learning*. London.

Hunter, W. 1768. Observations on the bones, commonly supposed to be elephants bones, which have been found near the River Ohio in America. *Phil. trans.* lviii, 34–45.

Hunter, W. 1770. Account of some bones, found in the rock of Gibraltar. *Phil. trans.* lx, 414–16.

Hutchinson, J. 1724–7. *Moses's Principia*. London.

Hutton, C., G. Shaw & R. Pearson (eds.) 1809. *Philosophical transactions from their commencement in 1665 to the year 1800, abridged*. 18 vols. London.

Hutton, J. 1785. *Abstract of a dissertation. . .concerning the system of the earth*. Edinburgh. Reprinted in V. A. Eyles (1970).

Hutton, J. 1788. Theory of the earth. *Trans. Roy. Soc. Edinburgh*. i, 209–304.

Hutton, J. 1792. *Dissertations on the different subjects in natural philosophy*. Edinburgh.

Hutton, J. 1794a. *A dissertation upon the philosophy of light, heat and fire*. Edinburgh.

Hutton, J. 1794b. *An investigation of the principles of knowledge*. 3 vols. Edinburgh.

Hutton, J. 1795. *The theory of the earth, with proofs and illustrations*. 2 vols. Edinburgh. 3rd vol. ed. by A. Geikie, London, 1899.

J.C. 1708. *The compleat collier*. London.

Jacob, M. C. 1969. John Toland and the Newtonian ideology. *Jnl of the Warburg and Courtauld Institutes*, xxxii, 307–31.

Jacob, M. C. 1971. The Church and the formulation of the Newtonian world-view. *Jnl of European studies*. i, 128–48.

Jacob, M. C. 1974. Early Newtonianism. *Hist. sci.* xii, 142–6.

Jacob, M. C. 1976. *The Newtonians and the English revolution*. London.

Jacob, M. C. & W. A. Lockwood. 1972. Political millenarianism and Burnet's *Sacred theory*. *Science studies*. ii, 265–79.

Jahn, M. E. 1962–8. The Old Ashmolean Museum and the Lhwyd Collections. *Jnl Soc. Bibliog. Nat. Hist*. iv, 244–8.

Jahn, M. E. 1969. Some notes on Dr Scheuchzer and on *Homo diluvii testis*. In C. J. Schneer (ed.), *Toward a history of geology*. Cambridge, Mass. 192–213.

Jahn, M. E. 1972. A bibliographical history of John Woodward's *An essay towards a natural history of the earth. Jnl Soc. Bibliog. Nat. Hist*. vi, 181–213.

Jaki, S. L. 1974. *Science and creation*. Edinburgh.

Jameson, R. 1795–96. Is the Volcanic opinion of the formation of Basaltes founded on truth? in Dissertations of the Royal Medical Society (MS), December 1795–96, vol. 35, no. xx, pp. 218–24; cited by J. M. Sweet & C. D. Waterston, 'Robert Jameson's approach to the Wernerian theory of the earth, 1796', *Ann. sci*. xxiii (1967), 81–95, p. 93.

Jameson, R. 1798. *An outline of the mineralogy of the Shetland Islands*. Edinburgh.

Jameson, R. 1800. *Mineralogy of the Scottish Isles*. 2 vols. Edinburgh.

Jameson, R. 1804–5. *System of mineralogy*. 2 vols. Edinburgh.

Jameson, R. 1805. *A treatise on the external character of minerals*. Edinburgh and London. Trans. from the German of A. G. Werner.

Jameson, R. 1808. *Treatise on geognosy*. Edinburgh (being the 3rd volume of the *System of mineralogy*).

Jameson, R. 1822. Introduction to *Cuvier's theory of the earth*. Trans. by R. Kerr. 4th ed. Edinburgh. 1st ed., 1813.

Jones F. W. 1953. John Hunter as a geologist. *Ann. of the Royal College of Surgeons*. xii, 219–44.

Jones, R. F. 1936. *Ancients and moderns*. St Louis.

Jones, W. B. 1871. *The Royal Institution*. London.

Jones, W. P. 1937. The vogue of natural history in England, 1750–70. *Ann. sci*. ii, 345–52.

Jones, W. P. 1966. *The rhetoric of science*. London.

Kargon, R. H. 1966. *Atomism in England from Harriot to Newton*. Oxford.

Kaufman, P. 1969. *Libraries and their users*. London.

Keill, J. 1698. *An examination of Dr. Burnet's Theory of the earth. Together with some remarks on Mr. Whiston's New theory of the earth*. Oxford.

Keill, J. 1699. *An examination of the Reflections on the theory* of the earth. *Together with a Defence of the remarks on Mr. Whiston's New theory.* Oxford.

Keill, J. 1720. *An introduction to natural philosophy.* London.

Keir, J. 1776. On the crystallizations observed on glass. *Phil. trans.* lxvi, 530–42.

Kendrick, T. D. 1956. *The Lisbon earthquake.* London.

Kidd, J. 1814. Notes on the mineralogy of the neighbourhood of St David's, Pembrokeshire. *Trans. Geol. Soc. London.* ii, 79–93.

Kidd, J. 1815. *A geological essay on the imperfect evidence in support of a theory of the earth.* Oxford.

King, W. 1685. A discourse concerning the bogs and Loughs of Ireland. *Phil. trans.* xv, 948–60.

Kircher, A. 1664–65. *Mundus subterraneus, in xii libros digestus.* Amsterdam.

Kirwan, R. 1791. Examination of the supposed igneous origin of stony substances. *Trans. Roy. Irish Acad.* v, 51–81.

Kirwan, R. 1794–6. *Elements of mineralogy.* 2 vols. London. 1st ed., 1784.

Kirwan, R. 1797. On the primitive state of the globe and its subsequent catastrophe. *Trans. Roy. Irish Acad.* vi, 233–308.

Kirwan, R. 1799. *Geological essays.* London.

Kirwan, R. 1802a. Observations on the proofs of the Huttonian theory of the earth. *Trans. Roy. Irish Acad.* viii, 3–27.

Kirwan, R. 1802b. An essay on the declivities of mountains. *Trans. Roy. Irish Acad.* viii, 35–52.

Kirwan, R. 1802c. Remarks on some skeptical positions in Mr Hume's Enquiry concerning the human understanding and his Treatise of human nature. *Trans. Roy. Irish Acad.* viii, 157–201.

Knight, D. M. 1970. The physical sciences and the Romantic movement. *Hist. sci.* ix, 54–75.

Kronick, D. A. 1962. *A history of scientific and technical periodicals, . . .1665–1790.* New York.

Kubrin, D. 1967. Newton and the cyclical cosmos: providence and the mechanical philosophy. *Jnl hist. ideas.* xxviii, 325–46.

Kubrin, D. C. 1968. Providence and the mechanical philosophy. The creation and dissolution of the world in Newtonian thought. A study of the relations of science and religion in seventeenth-century England. Ph.D. thesis. Cornell University.

Kuhn, A. J. 1961. Glory or gravity: Hutchinson vs Newton. *Jnl hist. ideas.* xxii, 303–22.

Kuhn, T. S. 1963. *The structure of scientific revolutions.* Chicago.

L.P. 1695. *Two essays sent from Oxford.* Oxford.

Lambarde, W. 1576. *A perambulation of Kent.* London.

Lancaster, W. T. (ed.) 1912. *Letters addressed to R. Thoresby, F.R.S.* Thoresby Society. Leeds.

Lang, W. D. 1935. Mary Anning of Lyme, collector and vendor of fossils, 1799–1847. *Natural history magazine.* v, 64–81.

Lankester, E. R. 1848. *The correspondence of John Ray.* Ray Society. London.

Leigh, C. 1700. *The natural history of Lancashire, Cheshire and the Peak.* Oxford.

Leland, J. 1745. *The itinerary of John Leland.* Ed. by T. Hearne. 9 vols. 2nd ed. Oxford.

Lhwyd, E. 1693. Epistola...de lapidum aliquot perpetua figura. *Phil. trans.* xvii, 746–54.

Lhwyd, E. 1699. *Lithophylacii Britannici ichnographia.* London.

Lhwyd, E. 1909. *Parochialia, Being a summary of answers to Parochial Queries in order to a geographical dictionary etc, of Wales.* London.

Lister, M. 1671. A letter of Mr. Martin Lister...adding some notes upon [the book of] M. Steno concerning petrify'd shells. *Phil. trans.* vi, 2281–5.

Lister, M. 1674. A description of certain stones figured like plants and by some observing men esteemed to be plants petrified. *Phil. trans.* vii, 6181–91.

Lister, M. 1674–5a. Some observations and experiments made...by Mr Martin Lister. *Phil. trans.* ix, 221–6.

Lister, M. 1674–5b. A letter of Mr Martin Lister, containing his observations of the Astroites or star-stones. *Phil. trans.* x, 274–9.

Lister, M. 1684a. Of the nature of earthquakes. *Phil. trans.* xiv, 512–17.

Lister, M. 1684b. An ingenious proposal for a new sort of maps of countreys. *Phil. trans.* xiv, 739–46.

Lister, M. A treatise on the fossils of England, chiefly the Northern parts (Bodleian Library, Lister MSS 7).

Lister, M. 1699. *A journey to Paris in the year 1698.* London.

Locke, J. 1693. *Some thoughts concerning education.* London.

Lough, J. 1970. *The Encyclopédie in eighteenth century England, and other studies.* Newcastle-upon-Tyne.

Lovell, R. 1661. *Panzoörykologia.* Oxford.

Lovejoy, A. O. 1936. *The great chain of being.* Cambridge, Mass.

Lovejoy, A. O. 1948. The parallel of Deism and Classicism. In Lovejoy, *Essays in the history of ideas.* Baltimore. 78–98.

Lowthorp, J. 1705. *Philosophical transactions and collections to 1700, abridged.* London.

Luc, J. A. de. 1779. *Lettres philosophiques et morales sur l'histoire de la terre et de l'homme.* 5 vols. La Haye.

Luc, J. A. de. 1790–1. Letters to Dr James Hutton, F.R.S., Edinburgh, on his *Theory of the earth. Monthly review.* n.s. ii, 1790, 206–27; 582–601; iii, 1790, 573–86; v, 1791, 564–85.

Luc, J. A. de. 1793–4. Geological letters addressed to Prof. Blumenbach. *British critic.* ii, 1793, 231–9; 351–8; iii, 1794, 110–20; 226–37; 467–78; 589–98.

Luc, J. A. de. 1809. *An elementary treatise on geology.* London. Trans. from French by H. de la Fite.

Luc, J. A. de. 1831. *Letters on the physical history of the earth, addressed to Professor Blumenbach.* To which are prefixed introductory remarks by H. de la Fite. London. (A reprinting, with additions, of his 1793–4 *British critic* letters.)

Lustkin, J. 1701. A letter about some large bones found in a gravel pit near Colchester. *Phil. trans.* xxii, 924–6.

Lyell, C. 1830–3. *Principles of geology.* 3 vols. London.

Lyell, Mrs K. M. 1881. *Sir Charles Lyell, Bart., Life, letters and journals.* Ed. by his sister-in-law. 2 vols. London.

Lyttleton, C. 1748. A beautiful Nautilites, shewn to the Royal Society. *Phil. trans.* xlv, 320–1.

MacCulloch, J. 1811a. Account of Guernsey, and the other Channel Islands. *Trans. Geol. Soc. London.* i, 1–22.

MacCulloch, J. 1811b. Notice accompanying a section of Heligoland. *Trans. Geol. Soc. London.* i, 322–3.

MacCulloch, J. 1819. *A description of the Western Islands of Scotland.* 3 vols. London.

McElroy, D. D. 1969. *Scotland's age of improvement.* Washington.

McGuire, J. E. & P. M. Rattansi. 1966. Newton and the 'pipes of Pan'. *Notes and records.* xxi, 108–43.

McIntyre, D. B. 1963. James Hutton and the philosophy of geology. In C. C. Albritton (ed.), *The fabric of geology.* Stanford. 1–11.

McKendrick, N. 1973. The role of science in the Industrial Revolution. A study of Josiah Wedgwood as a scientist and industrial chemist. In M. Teich & R. M. Young (eds.), *Changing perspectives in the history of science.* London. 274–319.

McKie, D. 1948. The scientific periodical from 1665–1798. *Phil. mag.* Commemorative ed. 122–32.

McKie, D. & E. Robinson. 1970. *Partners in science: letters of James Watt and Joseph Black.* London.

McMullin, E. 1970. The history and philosophy of science: a taxonomy. In R. H. Stuewer (ed.), *Minnesota studies in the philosophy of science.* v, 12–67.

Maillet, B. de. 1750. *Telliamed.* English trans. London.

Manuel, F. E. 1959. *The eighteenth century confronts the gods.* Cambridge, Mass.

Manuel, F. E. 1963. *Isaac Newton: historian.* Cambridge.

Manuel, F. E. 1968. *A portrait of Isaac Newton.* Cambridge, Mass.

Martin, B. 1735. *The philosophical grammar.* London.

Martin, W. 1793. *Figures and descriptions of petrifications collected in Derbyshire.* Wigan.

Martin, W. 1809a. *Petrificata Derbiensia.* Wigan.

Martin, W. 1809b. *Outlines of an attempt to establish a knowledge of extraneous fossils.* Macclesfield.

Mason, C. Manuscripts in Cambridge University Library (Add. MS 7762).

Matheson, C. 1954. Thomas Pennant and the Morris Brothers. *Ann. sci.* x, 258–71.

Maton, W. G. 1797. *Observations relative chiefly to the natural history, picturesque scenery, and the antiquities of the western counties of England, made in the years 1794 and 1796.* 2 vols. Salisbury.

Mendes da Costa, E. 1747. A dissertation on those fossil figured stones called Belemnites. *Phil. trans.* xliv, 397–407.

Mendes da Costa, E. 1749. A letter from Mr Emmanuel da Costa F.R.S. to the President concerning two beautiful Echinites. *Phil. trans.* xlvi, 143–8.

Mendes da Costa, E. 1757a. *A natural history of fossils.* Vol. i pt i. London.

Mendes da Costa, E. 1757b. An account of impressions of plants on the slates of coals. *Phil. trans.* l, 228–35.

Merret, C. 1667. *Pinax rerum naturalium Britannicarum.* London.

Merton, E. S. 1949. *Science and imagination in Sir T. Browne.* New York.

Merton, R. K. 1938. Science, technology and society in seventeenth century England. *Osiris.* iv, 360–632.

Merton, R. K. 1967. Science and technology in a democratic order. In Merton, *Social theory and social structure.* Glencoe. 550–61. Reprinted from the *Journal of legal and political sociology.* i, 1942.

Merz, J. T. 1896–1914. *A history of European thought in the nineteenth century.* 4 vols. Edinburgh.

Meteyard, E. 1871. *A group of Englishmen.* London.

Meyer, G. D. 1955. *The scientific lady in England, 1650–1760.* Berkeley.

Meyer, H. 1951. *The age of the world.* Allentown, Pa.

Michell, J. 1760. Conjectures concerning the cause and observations upon the phenomena of earthquakes. *Phil. trans.* li, 566–634.

Millar, J. 1810. Editor of 2nd ed. of J. Williams, *The natural history of the mineral kingdom.* Edinburgh.

Millhauser, M. 1954. The Scriptural geologists. *Osiris.* ii, 65–86.

Mills, A. 1790. Some account of the strata and volcanic appearances in the North of Ireland and Western Islands of Scotland. *Phil. trans.* lxxx, 73–100.

Mingay, G. E. 1970. *English landed society in the eighteenth century.* 3rd ed. London.

Mintz, S. I. 1962. *The hunting of Leviathan.* Cambridge.

Mitchell, Sir A. 1900–1. A list of travels, tours, journeys, cruises, excursions, wanderings, rambles, visits, etc. relating to Scotland. *Proc. of the Society of Antiquaries of Scotland.* xxxv, 431–638.

Mitchell, W. S. 1870–73. Notes on early geologists connected with the neighbourhood of Bath. *Proc. of the Bath Nat. Hist. and Antiquarian Field Club.* ii, 303–42.

Mitchison, R. M. 1962. *Agricultural Sir John.* London.

Moir, D. G. (ed.) 1973. *The early maps of Scotland, to 1850.* Vol. i, 3rd ed. Edinburgh.

Moir, E. 1964. *The discovery of Britain.* London.

Molyneux, T. 1694. Some notes upon the foregoing account of the Giants Causeway. *Phil. trans.* xviii, 175–82.

Molyneux, T. 1697. A discourse concerning the large horns frequently found under ground in Ireland. *Phil. trans.* xix, 489–512.

Molyneux, T. 1698. A letter containing some additional observations on the Giants Causeway in Ireland. *Phil. trans.* xx, 209–23.

Molyneux, T. 1700. An essay concerning giants. *Phil. trans.* xxii, 487–508.

Molyneux, W. 1683–4. Concerning Lough Neagh in Ireland and its petrifying qualitys. *Phil. trans.* xiii, 552–4.

Molyneux, W. 1697a. A correct draught of the Giants Causeway in Ireland. *Phil. trans.* xix, 333.

Molyneux, W. 1697b. A true description of the Bog of Kapinahane. *Phil. trans.* xix, 714–16.

Moore, W. 1851. *The Gentlemen's Society at Spalding: its origin and progress.* London.

Moray, Sir R. 1665a. A relation of persons killed with subterraneous damps. *Phil. trans.* i, 44–5.

Moray, Sir R. 1665b. An account how adits and mines are wrought at Liege without air-shafts. *Phil. trans.* i, 79–82.

More, S. 1782. An account of some scoriae from iron works. *Phil. trans.* lxxii, 50–2.

Mornet, D. 1911. *Les sciences de la nature au XVIII^e siècle.* Paris.

Morrell, J. B. 1971a. The University of Edinburgh in the late eighteenth century, its scientific eminence and academic structure. *Isis.* lxii, 158–71.

Morrell, J. B. 1971b. Professors Robison and Playfair and the *Theophobia Gallica. Notes and records.* xxvi, 43–63.

Morrell, J. B. 1972–3. Science and Scottish university reform: Edinburgh in 1826. *Brit. jnl hist. sci.* vi, 39–56.

Morrell, J. B. 1974. Reflections on the history of Scottish science. *Hist. sci.* xii, 81–94.

Morris, L. 1947. *Additional letters of the Morrises of Anglesey.* Ed. and transcribed by H. Owen. London.

Morton, J. 1712. *The natural history of Northampton-shire.* London.

Murray, Sir A. 1732. *An abstract of an essay on the improvement of husbandry and working of mines.* Edinburgh.

Murray, D. 1904. *Museums: their history and their use.* 3 vols. Glasgow.

[Murray, J.] 1802. *A comparative view of the Huttonian and Wernerian systems of geology.* Edinburgh.

Musson, A. E. & E. Robinson. 1969. *Science and technology in the Industrial Revolution.* Manchester.

Nangle, B. C. 1934 & 1955. *The Monthly review.* Oxford.

Nathanson, L. 1967. *The strategy of truth: a study of Sir Thomas Browne.* Chicago.

Needham, N. J. T. M. 1959. *Science and civilisation in China.* Vol. iii. London.

Nef, J. U. 1932. *The rise of the British coal industry.* 2 vols. London.

Neill, P. 1814. Biographical account of Mr Williams, the mineralogist. *Ann. phil.* iv, 81–3.

Nelson, C. M. 1968. Ammonites: Ammon's horns into cephalopods. *Jnl Soc. Bibliog. Nat. Hist.* v, 1–18.

Neuberg, V. E. 1971. *Popular education in eighteenth century England.* London.

Neve, M. & R. S. Porter. 1977. Alexander Catcott: Glory and geology. *Brit. jnl hist. sci.* x, 47–70.

Nichols, J. (ed.). 1809. *Letters on various subjects. . .to and from William Nicolson.* London.

Nichols, J. 1812–16. *Literary anecdotes of the eighteenth century.* 9 vols. London.

Nichols, J. 1817–58. *Illustrations of the literary history of the eighteenth century.* 8 vols. London.

Nicholson, N. 1955. *The Lakers.* London.

Nicols, T. 1652. *A lapidary.* Cambridge.

Nicolson, M. H. 1959. *Mountain gloom and mountain glory.* Ithaca, New York.

Nicolson, M. H. 1960. *The breaking of the circle.* New York.

Nixon, Rev. J. 1757. An account of the Temple of Serapis at Pozzuoli. *Phil. trans.* l, 166–74.

Nokes, D. L. 1975. A study of the Scriblerus Club. Ph.D. thesis. University of Cambridge.

North, F. J. 1928. *Geological maps, their history and development.* Cardiff.

North, F. J. 1934. 'The anatomy of the earth' – a seventeenth century cosmology. *Geological magazine.* lxxi, 541–7.

Oakley, K. P. 1965. Folklore of fossils. *Antiquity.* xxxix, 9–16; 117–125.

Oldenburg, H. 1675–6. Preface. *Phil. trans.* xi, 552.

Oldroyd, D. R. 1972. Robert Hooke's methodology of science as exemplified in his 'Discourse of earthquakes'. *Brit. jnl hist. sci.* vi, 109–30.

Oldroyd, D. R. 1974a. A note on the status of A. F. Cronstedt's simple earths and his analytical methods. *Isis.* lxv, 506–12.

Oldroyd, D. R. 1974b. Some phlogistic mineralogical schemes illustrative of the evolution of the concept of 'Earth' in the seventeenth and eighteenth centuries. *Ann. sci.* xxx, 269–306.

Oldroyd, D. R. 1975. Mineralogy and the 'chemical revolution'. *Centaurus.* xix, 54–71.

Oldroyd, D. R. & W. R. Albery. Forthcoming. The study of minerals and the order of things. *Brit. jnl hist. sci.*

Ornstein, M. 1913. *The role of the scientific societies in the seventeenth century.* New York.

Ospovat, A. M. 1969. Reflections on A. G. Werner's *Kurze Klassifikation.* In C. J. Schneer (ed.), *Toward a history of geology.* Cambridge, Mass. 242–56.

Otter, W. 1825. *The life and remains of Edward Daniel Clarke.* 2 vols. 2nd ed. London.

Owen, E. 1754. *Observations on the earth, rocks, stones and minerals for some miles about Bristol.* London.

Owen, R. 1861. *John Hunter, F.R.S. Essays and observations on natural history, anatomy, physiology, psychology and geology.* 2 vols. London.

Paffard, M. 1970. Robert Plot – a county historian. *History today.* xx, 112–17.

Page, L. E. 1963. The rise of diluvial theory in British geological thought. Ph.D thesis. University of Oklahoma.

Page, L. E. 1969. Diluvialism and its critics in Great Britain in the early nineteenth century. In C. J. Schneer (ed.), *Toward a history of geology*. Cambridge. Mass. 257–71.

Paley, W. 1802. *Natural theology*. London.

Pantin, C. F. A. 1968. *The relations between the sciences*. Cambridge.

Paris, J. A. 1818. Observations on the geological structure of Cornwall. *Trans. Roy. Geol. Soc. Cornwall*. i, 168–200.

Parkes, J. 1925. *Travel in England in the seventeenth century*. London.

Parkinson, J. 1804–11. *Organic remains of a former world*. 3 vols. London.

Parkinson, J. 1811. Observations on some of the strata in the neighbourhood of London, and on the fossil remains contained in them. *Trans. Geol. Soc. London*. i, 324–54.

Parris, L. 1973. *Landscape in Britain, c. 1750–1850*. London.

Parsons, Dr J. 1755. Remarks upon a petrified Echinus. *Phil. trans.* xlix, 155–6.

Parsons, Dr J. 1757. An account of some fossile fruits, and other bodies found in the Island of Sheppey. *Phil. trans.* l, 396–407.

Pattison, M. 1859. Tendencies of religious thought in England, 1688–1750. In *Essays and reviews*. London.

Pennant, T. 1771. *A tour in Scotland*. Chester.

Pennant, T. 1810. *Tours in Wales*. 3 vols. London. 1st ed., 1772.

Pettus, Sir J. 1686. *Fleta minor, or the laws of art and nature in knowing, judging, assaying, fining, refining, and inlarging the bodies of confin'd metals*. London.

Phillips, J. 1839. Biographical notice of William Smith, LL.D. *Magazine of natural history*. iii, 213–20.

Phillips, J. 1844. *Memoirs of William Smith*. London.

Phillips, W. 1815. *An outline of mineralogy and geology*. London.

Phillipson, N. T. & R. Mitchison (eds.) 1970. *Scotland in the age of improvement*. Edinburgh.

Phillipson, N. T. 1973. Towards a definition of the Scottish Enlightenment. In P. Fritz & D. Williams (eds.), *City and society in the eighteenth century*, Toronto. 125–47.

*Phil. trans.* 1666a. General heads for a natural history of a countrey. *Phil. trans.* i, 186–9.

*Phil. trans.* 1666b. Observables touching petrification. *Phil. trans.* i, 320–1.

*Phil. trans.* 1666c. Articles of inquiries touching mines. *Phil. trans.* i, 330–43.

*Phil. trans.* 1673. Directions for inquiries concerning stones and other materials for the use of buildings. *Phil. trans.* viii, 6015.

Piggott, S. 1950. *William Stukeley*. Oxford.

Piggott, S. 1968. *The Druids*. London.

Pigot, T. 1683. An account of the earthquake that happened at Oxford. *Phil. trans.* xiii, 311–21.

Pilkington, J. 1789. *A view of the present state of Derbyshire*. 2 vols. Derby.

Pinkerton, J. 1811. *Petrology: a treatise on rocks*. London.

Piper, H. W. 1962. *The active universe*. London.

Pitman, J. H. 1924. *Goldsmith's animated nature*. New Haven.

Platt, J. 1750. A letter...concerning a flat spheroidal stone. *Phil. trans.* xlvi, 534–5.

Platt, J. 1758. An account of the fossile thigh-bone of a large animal dug up at Stonesfield, near Woodstock, Oxford. *Phil. trans.* l, 524–527.

Plattes, G. 1639. *A discovery of subterraneall treasure*. London.

Playfair, J. 1802. *Illustrations of the Huttonian theory of the earth*. Edinburgh.

Playfair, J. 1805. Biographical account of the late Dr James Hutton. *Trans. Roy. Soc. Edinburgh.* v, 39–99.

[Playfair, J.] 1811. Review of *Trans. Geol. Soc. London. Edin. rev.* xix, 207–29.

[Playfair, J.] 1814. Review of *Essay on the Theory of the earth*, trans. from the French of M. Cuvier. *Edin. rev.* xliv, 454–74.

Plot, R. 1677. *The natural history of Oxford-shire*. Oxford.

Plot, R. 1686. *The natural history of Stafford-shire*. Oxford.

Plumb, J. H. 1964. The historian's dilemma. In J. H. Plumb (ed.), *Crisis in the humanities*. Harmondsworth. 24–44.

Plumb, J. H. 1967. *The growth of political stability in England, 1675–1725*. London.

Plumb, J. H. 1972. The public, literature and the arts in the eighteenth century. In P. Fritz & D. Williams (eds.), *The triumph of culture*. Toronto. 27–48.

Plumb, J. H. 1973. *The commercialization of leisure in eighteenth century England*. Reading.

Plymley, Rev. J. 1803. *General view of the agriculture of Shropshire*. London.

Pococke, R. 1744. *An account of the glaciers or Ice Alps in Savoy*. London.

Pococke, R. 1748. An account of the Giants Causeway in Ireland. *Phil. trans.* xlv, 124–7.

Pococke, R. 1753. A farther account of the Giants Causeway. *Phil. trans.* xlviii, 226–37.

Pococke, R. 1888. *Travels through England during 1750, 1751 and later years*. Ed. by J. J. Cartwright. London.

Pococke, R. 1891. *Tour of Ireland in 1752.* Ed. with notes by G. T. Stokes. Dublin.

Pollard, S. 1965. *The genesis of scientific management.* London.

Porter, R. S. 1972. Review of *Science in the British colonies of America* by R. P. Stearns. *Historical jnl.* xv, 175–7.

Porter, R. S. 1973a. Review of *Charles Lyell, the revolution in geology* by L. J. Wilson. *Jnl of modern history.* xlv, 503–4.

Porter, R. S. 1973b. The industrial revolution and the rise of the science of geology. In M. Teich & R. M. Young (eds.), *Changing perspectives in the history of science.* London. 320–43.

Porter, R. S. 1974. The making of the science of geology in Britain, 1660–1815. Ph.D. thesis. University of Cambridge.

Porter, R. S. 1976a. Charles Lyell and the principles of the history of geology. *Brit. jnl hist. sci.* ix, 91–103.

Porter, R. S. 1976b. William Hobbs of Weymouth and his *The earth generated and anatomized. Jnl Soc. Bibliog. Nat. Hist.* vii, 333–41.

Proctor, I. R. 1973. The concept of granite in the late eighteenth century and early nineteenth century in relation to the theories of the earth. M.Litt. thesis. University of Cambridge.

Pryce, W. 1778. *Mineralogia Cornubiensis.* London.

Pryme, A. de la. 1700. A letter. . .concerning Broughton in Lincolnshire with observations on the shell-fish observed in the quarries about that place. *Phil. trans.* xxii, 677–87.

Pryme, A. de la. 1869. *The diary of Abraham de la Pryme.* Ripon.

Quarrell, W. H. & W. J. C. (eds.) 1928. *Oxford in 1710, from the travels of Z. C. von Uffenbach.* Oxford.

Quinlan, M. J. 1941. *Victorian prelude.* New York.

Raistrick, A. 1938. *Two centuries of industrial welfare. The London Quaker Lead Company, 1692–1905.* London.

Raistrick, A. 1968. *Quakers in science and industry.* Newton Abbot. 1st ed., 1950.

Ramsay, J., of Ochtertyre. 1966. *The letters of John Ramsay of Ochtertyre.* Ed. by B. L. H. Horn. Edinburgh.

Rappaport, R. 1969. The geological atlas of Guettard, Lavoisier, and Monnet: conflicting views on the nature of geology. In C. J. Schneer (ed.), *Toward a history of geology.* Cambridge, Mass. 272–87.

Raspe, R. E. 1771. A letter containing a short account of some Basalt hills in Hassia. *Phil. trans.* lxi, 580–3.

Raspe, R. E. 1776. *Johann Jakob Ferber: travels through Italy in the years 1771 and 1772.* Trans. from the German by R. E. Raspe. London.

Raspe, R. E. 1777. *Ignaz Edler von Born: Travels through the Bannat of Temeswar, Transylvania and Hungary in the year 1770.* Trans. by R. E. Raspe. London.

Raspe, R. E. 1970. *Specimen historiae naturalis globi terraquei.* Trans. and ed. by A. V. Carozzi. New York. 1st ed., 1763.

Rattansi, P. M. 1972. The social interpretation of seventeenth century science. In P. Mathias (ed.), *Science and society, 1600–1900.* London. 1–33.

Raven, C. E. 1947. *English naturalists from Neckam to Ray.* Cambridge.

Raven, C. E. 1950. *John Ray.* 2nd ed. Cambridge.

Raven, C. E. 1954. *Organic design: a study of scientific thought from Ray to Paley.* London.

Ravetz, J. R. 1971. *Scientific knowledge and its social problems.* Oxford.

Ray, J. 1673. *Observations topographical, moral and physiological.* London. 2nd ed., 1738.

Ray, J. 1691. *The wisdom of God manifested in the works of the Creation.* London. 3rd ed., 1701.

Ray, J. 1692. *Miscellaneous discourses concerning the dissolution and changes of the world.* London.

Ray, J. 1693. *Three physico-theological discourses.* 2nd ed. London. 4th ed., 1732.

Read, D. 1964. *The English provinces, c. 1760–1960.* London.

Redwood, J. 1976. *Reason, ridicule and religion: The age of Enlightenment in England, 1660–1750.* London.

Rennie, J. 1816. 'A comparative view of the Huttonian and the Wernerian theories of the earth' (MS in National Library of Scotland, 2726).

Rhodes, J. N. 1968. Dr Linden, William Hooson, and North Welsh mining in the mid eighteenth century. *Bulletin of the Peak District Mines History Society.* iii, 259–70.

Richardson, R. 1697. Part of a letter...containing a relation of subterraneous trees dug up at Youle in Yorkshire. *Phil. trans.* xix, 526–8.

Richardson, R. 1835. *Extracts from the literary and scientific correspondence of Richard Richardson.* Yarmouth.

Richardson, W. 1808. A letter on the alterations that have taken place in the structure of rocks. *Phil. trans.* xcviii, 187–222.

Richardson, W. 1811. On the strata of mountains. *Phil. mag.* xxxvii, 367–9.

Ritchie, J. 1952. Natural history and the emergence of geology in the Scottish universities. *Trans. of the Edinburgh Geological Society.* xv, 297–316.

Ritterbush, P. C. 1964. *Overtures to biology.* New Haven.

Robertson, A. 1922. *The life of Sir Robert Moray*. London.

Robinson, T. 1694. *The anatomy of the earth*. London.

Robinson, T. 1696. *New observations on the natural history of this world of matter and this world of life*. London.

Robinson, T. 1709a. *An essay towards a natural history of Westmorland and Cumberland*. London.

Robinson, T. 1709b. *A vindication of the philosophical and theological exposition of the Mosaick system of the Creation*. London.

Roger, J. 1973–4. La théorie de la terre au XVII<sup>e</sup> siècle. *Revue d'histoire des sciences*. xxvi, 23–48.

Ronan, C. 1970. *Edmond Halley: genius in eclipse*. London.

Rosenberg, C. E. 1970. On writing the history of American science. In H. J. Bass (ed.), *The state of American history*. Chicago.

Rostow, W. W. 1960. *The stages of economic growth*. Cambridge.

Rousseau, G. S. 1968. The London earthquakes of 1750. *Cahiers d'histoire mondiale*. xi, 436–51.

Royal Society of Edinburgh. Minute Book of the Meetings of the Physical and Literary Classes of the Royal Society of Edinburgh 1 July 1793 to 12 January 1824 (MS in Royal Society of Edinburgh).

Rudel, A. 1962. *Les volcans d'Auvergne*. Clermont-Ferrand.

Rudwick, M. J. S. 1962. Hutton and Werner compared: George Bellas Greenough's geological tour of Scotland in 1805. *Brit. jnl hist. sci.* i, 117–35.

Rudwick, M. J. S. 1963. The foundation of the Geological Society of London. *Brit. jnl hist. sci.* i, 325–55.

Rudwick, M. J. S. 1967. A critique of Uniformitarian geology. *Proc. Amer. Phil. Soc.* cxi, 272–87.

Rudwick, M. J. S. 1971. Uniformity and progression: reflections on the structure of geological theory in the age of Lyell. In D. H. D. Roller (ed.), *Perspectives in the history of science and technology*. Norman, Oklahoma.

Rudwick, M. J. S. 1972. *The meaning of fossils*. London.

Rudwick, M. J. S. 1974. Poulett Scrope on the volcanoes of Auvergne. *Brit. jnl hist. sci.* vii, 205–42.

Rudwick, M. J. S. 1975. *The history of the natural sciences as cultural history*. Inaugural lecture, Vrije Universiteit, Amsterdam.

Rudwick, M. J. S. 1976. The emergence of a visual language for geological science, 1760–1840. *Hist. sci.* xiv, 149–95.

Ruskin, J. 1903–12. *Works*. 39 vols. London.

St Clair, C. 1966. The classification of minerals: some representative systems from Agricola to Werner. Ph.D. thesis. University of Oklahoma.

Saussure, H. B. de. 1799. Agenda, or a collection of observations and researches. *Phil. mag.* iii, 33–41; 147–56; 194–99; iv, 68–71; v, 24–29; 135–40; 217–21.

Schall, C. F. W. 1787. *Oryctologische Bibliothek.* Weimar.

Scherz, G. (ed.) 1956. *Vom Wege Niels Stensens.* Copenhagen.

Schilling, B. N. 1950. *Conservative England and the case against Voltaire.* New York.

Schmitt, C. B. 1973. Toward a reassessment of Renaissance Aristotelianism. *Hist. sci.* xi, 159–93.

Schneer, C. J. 1954. The rise of historical geology in the seventeenth century. *Isis.* xlv, 256–68.

Schneer, C. J. (ed.) 1969. *Toward a history of geology.* Cambridge, Mass.

Schofield, R. E. 1963a. Histories of scientific societies. *Hist. sci.* ii, 70–83.

Schofield, R. E. 1963b. *The Lunar Society of Birmingham.* Oxford.

Schofield, R. E. 1970. *Mechanism and materalism.* Princeton.

Scilla, A. 1696. Review of *La vana speculatione. Phil. trans.* xix, 181–95.

Scott, E. J. L. 1904. *Index to the Sloane Manuscripts in the British Museum.* London.

Scott, H. W. (ed.) 1966. *John Walker: Lectures on geology, including hydrography, mineralogy and meteorology.* Notes and introduction by H. W. Scott. Chicago.

Sedgwick, A. 1831. Presidential address to the Geological Society of London, 1831. *Proc. Geol. Soc. London.* xx, 270–316.

Septalius, S. M. 1667. Some observations communicated...concerning quicksilver. *Phil. trans.* ii, 493.

Sewell, E. 1961. *The orphic voice: poetry and natural history.* London.

Shapin, S. 1971. Hanoverian science in its cultural context: the Royal Society of Edinburgh. Ph.D. thesis. University of Pennsylvania.

Shapin, S. 1974a. Property, patronage and the politics of science. *Brit. jnl hist. sci.* vii, 1–41.

Shapin, S. 1974b. The audience for science in eighteenth century Edinburgh. *Hist. sci.* xii, 95–121.

Shapin, S. & A. W. Thackray. 1974. Prosopography as a research tool in history of science: the British scientific community 1700–1900. *Hist. sci.* xii, 1–28.

Sharp, L. W. (ed.) 1937. *Early letters of Robert Wodrow, 1698–1709.* Scottish History Society. Edinburgh.

[Sharpe, W.] 1769. *Treatise upon coal mines.* London.

Shaw, S. 1798–1801. *The history and antiquities of Staffordshire.* 2 vols. London.

Sheppard, T. 1914–22. William Smith, his maps and memoirs. *Yorkshire Geol. Soc. Proc.* xix, 75–253.

Sherley, T. 1672. *A philosophical essay declaring the probable causes whence stones are produced in the greater world.* London.

Sibbald, Sir R. 1684. *Scotia illustrata.* Edinburgh.

Sibbald, Sir R. 1697. *Auctarium Musaei Balfouriani e Museo Sibbaldiano.* Edinburgh.

Sinclair, G. 1672. *The hydrostaticks.* Edinburgh.

Sisco, A. G. & C. S. Smith. 1951. *Lazarus Ercker's Treatise on ores and assaying.* Chicago.

Skelton, R. A. 1970. *County atlases of the British Isles 1579–1703.* London.

Skinner, Q. R. D. 1969. Meaning and understanding in the history of ideas. *History and theory.* viii, 3–53.

Sleep, M. C. W. 1969. Sir William Hamilton (1730–1803), his work and influence in geology. *Ann. sci.* xxv, 319–38.

Sloane, H. 1693. Several regularly figured stones. *Phil. trans.* xvii, 188.

Smellie, W. 1790–9. *The philosophy of natural history.* 2 vols. Edinburgh.

Smith, A. 1795. *Essays on philosophical subjects.* Ed. by J. Black & J. Hutton, London.

Smith, C. S. 1969. Porcelain and Plutonism. In C. J. Schneer (ed.), *Toward a history of geology.* Cambridge, Mass. 317–38.

Smith, S. 1943. Seventeenth century observations on rocks, minerals and fossils etc. of Gloucestershire and Somerset. *Proc. of the Bristol Naturalists Society.* 4th series. ix, 406–26.

Smith, S. 1946. Owen's observations. *Proc. of the Bristol Naturalists Society.* xxvii, 93–103.

Smith, Rev. W. 1745. *A natural history of Nevis.* Cambridge.

Smith, W. C. 1911–13. The mineral collection of Thomas Pennant, 1726–1798. *The mineralogical magazine.* xvi, 331–42.

Smith, W. C. 1969. A history of the first hundred years of the Mineral Collections in the British Museum. *Bulletin of the British Museum (Natural History) historical series.* iii, 235–59.

Smith, W. C. 1976. The Mineralogical Society, 1876–1976. *The mineralogical magazine.* xl, 429–39.

Smith, W. 1815. *A memoir to the map and delineation of the strata of England and Wales with part of Scotland.* London.

Smith, W. 1816. *Strata identified by organized fossils.* London.

Southwell, Sir R. 1683. A description of Pen Park Hole. *Phil. trans.* xiii, 2–6.

Stearns, R. P. 1970. *Science in the British colonies of America.* Urbana.

Steno, N. 1669. *De solido intra solidum naturaliter contento dissertationis prodromus.* Florence. Trans. into English by Henry Oldenburg, London, 1671.

Stephen, L. 1876. *History of English thought in the eighteenth century.* London.

Stillingfleet, E. 1662. *Origines sacrae.* London.

Stimson, D. 1948. *Scientists and amateurs: a history of the Royal Society.* New York.

Stock, J. E. 1811. *Memoir of the life of Thomas Beddoes, M.D.* London.

Stoye, J. 1952. *English travellers abroad. 1604–67.* London.

Strachey, J. 1719. A curious description of the strata observ'd in the coal-mines of Mendip in Somersetshire. *Phil. trans.* xxx, 968–73.

Strachey, J. 1725. An account of the strata in coal-mines. *Phil. trans.* xxxiii, 395–8.

Strange, Sir J. 1775a. An account of Two Giants Causeways, or groups of prismatic basaltine columns, and other curious vulcanic concretions. *Phil. trans.* lxv, 5–47.

Strange, Sir J. 1775b. An account of a curious Giants Causeway, or group of angular columns, newly discovered in the Euganean Hills. *Phil. trans.* lxv, 418–23.

Stukeley, W. 1724. *Itinerarium curiosum.* London.

Stukeley, W. 1750. On the causes of earthquakes. *Phil. trans.* lxvi, 641–6.

Stukeley, W. 1882–7. *The family memoirs of the Rev. William Stukeley...and other correspondence.* Ed. by W. C. Lukis. 3 vols. Durham.

Sulivan, Sir R. J. 1794. *A view of nature.* 6 vols. London.

Sweet, J. M. 1935. The mineral collections of Sir Hans Sloane. *Natural history magazine.* v, 49–64; 97–116; 145–64.

Sweet, J. M. 1963. Robert Jameson in London, 1793. *Ann. sci.* xix, 81–116.

Sweet, J. M. 1967. Robert Jameson's Irish journal, 1797. *Ann. sci.* xxiii, 97–126.

Sweet, J. M. 1972. Instructions to collectors – John Walker (1793) and Robert Jameson (1817); with biographical notes. *Ann. sci.* xxix, 397–414.

Sweet, J. M. & C. Waterston. 1967. Robert Jameson's approach to the Wernerian theory of the earth, 1796. *Ann. sci.* xxiii, 81–95.

Swift, J. 1906. *Gulliver's travels.* Everyman ed. London. 1st ed., 1726.

Taylor, E. G. R. 1934. *Late Tudor and early Stuart geography, 1583–1650.* London.

Taylor, S. 1730. *The history and antiquities of Harwich and Dovercourt.* London.

Tentzelius, W. E. 1697. Epistola de sceleto elephantino. *Phil. trans.* xix, 757–76.

Thackray, A. W. 1970a. *Atoms and powers.* Cambridge, Mass.

Thackray, A. W. 1970b. Science: has its present past a future? In R. H. Stuewer (ed.), *Minnesota studies in the philosophy of science.* v, 112–26.

Thackray, A. W. 1974. Natural knowledge in cultural context: the Manchester model. *American historical review.* lxxix, 672–709.

Thomas, K. 1971. *Religion and the decline of magic.* London.

Thompson, E. P. 1968. *The making of the English working class.* Harmondsworth.

Tillyard, E. W. 1943. *The Elizabethan world-picture.* London.

Todd, A. C. 1955–64. Origins of the Royal Geological Society of Cornwall. *Trans. Roy. Geol. Soc. Cornwall.* xix, 179–84.

Tomkeieff, S. I. 1947–9. James Hutton and the philosophy of geology. *Proc. Roy. Soc. Edinburgh.* lxiii, 387–400.

Tomkeieff, S. I. 1962. Unconformity – an historical study. *Proc. Geol. Assoc.* lxxiii, 383–417.

Torrens, H. S. 1974a. William Sharpe and the search for coal near Sherborne. *The Shirburnian.* June. 22–5.

Torrens, H. S. 1974b. Lichfield museums (pre 1850). *Newsletter of the Geological Curators Group.* i, 5–9.

Torrens, H. S. 1974c. Lichfield Museums (pre 1850): a postscript. *Newsletter of the Geological Curators Group.* ii, 38–9.

Torrey, N. L. 1930. *Voltaire and the English Deists.* New Haven.

Toulmin, G. H. 1785. *The eternity of the world.* London.

Toulmin, G. H. 1789. *The eternity of the universe.* London.

Toulmin, S. E. 1972. *Human understanding.* London.

Toulmin, S. E. & J. Goodfield. 1967. *The discovery of time.* Harmondsworth.

Townsend, Rev. J. 1813–15. *The character of Moses established for veracity as an historian.* 2 vols. Bath.

Townson, R. 1797. *Travels in Hungary.* London.

Townson, R. 1798. *Philosophy of mineralogy.* London.

Tredgold, T. 1818. Remarks on the geological principles of Werner and those of Mr Smith. *Phil. mag.* li, 36–8.

Treneer, A. 1963. *The mercurial chemist: a life of Sir H. Davy.* London.

Tuan, Yi-fu. 1968. *The hydrologic cycle and the wisdom of God.* Toronto.

Tunbridge, P. A. 1971. Jean André de Luc, F.R.S. (1727–1817). *Notes and records.* xxvi, 15–33.

Turnbull, H. W. (ed.) 1959. *The correspondence of Isaac Newton.* Vol. ii. Cambridge.

Turner, A. J. 1974. Hooke's theory of the earth's axial displacement. *Brit. jnl hist. sci.* vii, 166–70.

Turner, Rev. W. 1802. *A general introductory discourse on the objects, advantages and intended plan of the New Institution for Public Lectures, on natural philosophy in Newcastle-upon-Tyne.* Newcastle-upon-Tyne.

Tuveson, E. L. 1949. *Millennium and Utopia.* Berkeley.

Tuveson, E. L. 1950. Swift and the world-makers. *Jnl hist. ideas.* xi, 54–74.

Upcott, W. 1818. *A bibliographical account of the principal works relative to English topography.* 3 vols. London.

Vereker, C. H. 1967. *Eighteenth century optimism.* Liverpool.

Vivian, J. H. 1815a & b. Two letters to Robert Jameson, 5 Oct. and 24 Oct. (MSS in Edinburgh University Library, MS JA 9).

Vivian, J. H. 1818. A sketch of the plan of the Mining Academies of Freyberg and Schemnitz. *Trans. Roy. Geol. Soc. Cornwall.* i, 71–7.

Walcott, J. 1779. *Descriptions and figures of petrifactions found in the quarries, gravel pits, etc. near Bath.* Bath.

Walker, D. P. 1972. *The ancient theology.* London.

Walker, J. 1764. Letter to Lord Kames, 10 Dec. (Edinburgh University Library, La III 352).

Walker, J. 1789. Letters to Sir W. Pulteney, 27 March (Edinburgh University Library, La III 352).

Walker, J. Occasional remarks (MS in Edinburgh University Library, DC 2 40).

Walker, J. 1808. *An economical history of the Hebrides and Highlands of Scotland.* 2 vols. Edinburgh.

Wallis, J. 1701. A letter. . .relating to that Isthmus or neck of land which is supposed to have joyned England and France in former times. *Phil. trans.* xxii, 967–79; 1030–8.

Warburton, H. 1814. Annual report of the Council of the Geological Society of London. Printed report, of which there is a copy in the Greenough MSS (Cambridge University Library).

Ward, J. 1740. *Lives of the professors of Gresham College.* London.

Warren, E. 1690. *Geologia, or a discourse concerning the earth before the Deluge.* London.

Waterston, C. 1965. William Thomson 1761–1806: a forgotten benefactor. *University of Edinburgh Journal.* Autumn. 120–34.

Watson, R. 1806. *Two apologies, one for Christianity...the other for the Bible.* London.

Watson, R. S. 1897. *The history of the Literary and Philosophical Society of Newcastle-upon-Tyne, 1793–1896.* London.

Watt, G. 1804. Observations on basalt. *Phil. trans.* 279–314.

Webster, C. 1967. The origins of the Royal Society. *Hist. sci.* vi, 106–28.

Webster, C. 1974. *The intellectual revolution of the seventeenth century.* London.

Webster, C. 1975. *The great instauration.* London.

Webster, J. 1671. *Metallographa.* London.

Wedgwood, J. 1783. Some experiments upon the ochra friabilis nigro fusca of Da Costa etc. *Phil. trans.* lxxiii, 284–7.

Wedgwood, J. 1790. On the analysis of a mineral substance from New South Wales. *Phil. trans.* lxxx, 306–20.

Weld, C. R. 1848. *A history of the Royal Society.* 2 vols. London.

Werner, A. G. 1774. *Von den äusserlichen Kennzeichen der Fossilien.* Leipzig.

Werner, A. G. 1971. *Short classification and description of the various rocks.* Trans. and with an introduction by A. Ospovat. New York.

West, T. 1777. An account of a volcanic hill near Inverness. *Phil. trans.* lxvii, 385–7.

Westfall, R. S. 1973. *Science and religion in seventeenth century England.* 2nd ed. Ann Arbor.

Whalley, P. E. S. 1971. William Arderon, F.R.S., of Norwich, an eighteenth century diarist and letter writer. *Jnl Soc. Bibliog. Nat. Hist.* vi, 30–49.

Whewell, W. 1837. *History of the inductive sciences from the earliest to the present times.* 3 vols. London.

Whiston, W. 1696. *A new theory of the earth.* London.

Whiston, W. 1698. *A vindication of the New theory of the earth from the exceptions of Mr Keill and others.* London.

Whiston, W. 1700. *A second defence of the New theory of the earth from the exceptions of Mr John Keill.* London.

Whiston, W. 1714. *The cause of the Deluge demonstrated.* London.

Whiston, W. 1753. *Memoirs of the life and writings of Mr W. Whiston.* 2 vols. London. 1st ed., 1749.

White, A. D. 1876. *The warfare of science with theology in Christendom.* London.

Whitehurst, J. 1778. *An inquiry into the original state and formation of the earth.* London.

Willey, B. 1934. *The seventeenth century background*. London.

Willey, B. 1940. *The eighteenth century background*. London.

Williams, A. 1948. *The common expositor: an account of the commentaries on Genesis, 1527–1633*. Chapel Hill, N.C.

Williams, J. 1787. *An account of some remarkable ancient ruins*. London.

Williams, J. 1789. *The natural history of the mineral kingdom*. 2 vols. Edinburgh.

Williamson, G. 1935. Mutability, decay and seventeenth century melancholy. *English Literary History*. ii, 121–50.

Wilson, L. G. 1972. *Charles Lyell: the years to 1841: the revolution in geology*. Vol. i. New Haven.

Winch, N. 1817. Observations on the geology of Northumberland and Durham. *Trans. Geol. Soc. London*. iv, 1–101.

Winter, H. J. J. 1939. Scientific notes from the early minutes of the Peterborough Society, 1730–45. *Isis*. xxxi, 51–9.

Winter, H. J. J. 1950. Scientific associations of the Spalding Gentleman's Society, 1710–50. *Archives internationales d'histoire des sciences*. 77–88.

Withering, W. 1782. An analysis of two mineral substances, viz the Rowley Rag-stone and the Toad-stone. *Phil. trans.* lxxii, 327–36.

Wittlin, A. S. 1949. *The museum, its history and its tasks in education*. London.

Wollaston, W. 1811. Letter of 25 Sept. 1811 to T. Allan (National Library of Scotland, MS 583 no. 749).

Wood, Sir H. T. 1913. *A history of the Royal Society of Arts*. London.

Woodward, H. B. 1907. *The history of the Geological Society of London*. London.

Woodward, J. 1695. *An essay toward a natural history of the earth*. London.

Woodward, J. 1696. *Brief instructions for making observations in all parts of the world*. London.

Woodward, J. 1714. *Methodica et ad ipsam naturae normam instituta fossilium in classes distributio*. London.

Woodward, J. 1726. *The natural history of the earth illustrated, inlarged and defended*. London.

Woodward, J. 1728. *Fossils of all kinds digested into a method.* · London.

Woodward, J. 1729. *An attempt towards a natural history of the fossils of England*. London.

Woodward, J. MS Gough Wales 8 (Bodleian Library, Oxford).

Worthington, W. 1773. *The scripture theory of the earth*. London.

Wotton, W. 1697. *A vindication of an abstract of an Italian book*.

Printed in J. Arbuthnot, *An examination of Dr Woodward's account of the Deluge.* London.

Wright, T. 1668. A curious and exact relation of a sand-flood. *Phil. trans.* iv, 722–5.

Yates, F. A. 1964. *Giordano Bruno and the Hermetic tradition.* London.

Young, R. M. 1973. The historiographic and ideological contexts of the nineteenth century debate on man's place in nature. In M. Teich & R. M. Young (eds.), *Changing perspectives in the history of science.* London.

Zeman, F. D. 1950. William Withering as a mineralogist. *Bulletin of the history of medicine.* xxiv, 530–8.

Zittel, K. A. 1901. *History of geology and palaeontology.* Trans. and abridged by M. Ogilvie-Gordon. London. Reprinted, Weinheim, 1962.

Zollman, P. H. 1746. An extract by Philip Henry Zollman FRS of... an Italian book, intitled *De crustacei ed altri marini corpi che se trovano sui monti* di Antoni Lazzaro Moro, Venice, 1740. *Phil. trans.* xliv, 163–6.

# Index